Hayek: a commemorative album

Hayek
a commemorative album

compiled by
John Raybould

ADAM SMITH INSTITUTE
LONDON

Published by the Adam Smith Institute,
21 Great Smith Street, London SW1P 3BL, UK
Tel: +44 (0)171 222 4995
Fax: +44 (0)171 222 7544

ISBN: 1-873712-95-2

Designed, copy-edited and typeset by
Sukie Hunter, The Smithy House, Stichill,
Roxburghshire, UK

Printed in Great Britain by
Hartington Fine Arts Ltd, Lancing, West Sussex

Foreword: Hayek's legacy

F. A. Hayek was the 20th century's leading philosopher of liberty.

By the time he received the Nobel Prize for Economics in 1974 at the age of 75, he had written 25 books and more than 130 articles – not just on economics, but on political and legal philosophy, intellectual history and even psychology.

His 1944 book, *The Road to Serfdom*, was perhaps the greatest source of his lifetime influence. Arguing that socialism, however modest in intent, would inevitably produce tyranny in the end, it was an instant sensation, read by millions within months of publication.

Yet Hayek's most lasting legacy is probably his demonstration of how societies can flourish without the need for conscious planning: indeed, they fare better in its absence. For societies are spontaneous orders – formed through human action right enough, but not the products of human design. Like language, they grow and develop, without needing anyone to command them to.

Thus the price mechanism co-ordinates the activities of millions upon millions of individuals. Rising prices indicate an excess of demand over supply, prompting individuals in every part of the globe to divert their efforts into meeting that demand – as if by magic, and without their needing to be told to change.

This market exchange system is vast, complex and subtle. That is why co-ordination breaks down when we try to manipulate the system or to replace it with some supposedly more rational alternative. Our reason alone is simply not up to the task. Government should aim not to impose its own preconceived order on society but to remove obstacles to the free operation of the market and to discover the rules of justice that make the spontaneous social order work.

In short, Hayek has left us with a powerful and positive case for free-market capitalism and for the free society that will influence all future debate on the subject.

Yet he was also a man – a scholarly, modest, rather shy man, who had an instinctive flair for promoting liberal ideas, and yet seemed slightly overwhelmed by his successes in doing so. It is this human side of Hayek that John Raybould brings out in this book: drawing on family and other archives to illustrate the life of a remarkable scholar, and a remarkable man.

Eamonn Butler

Preface

This book uses original photographs and contemporary documents to illustrate the life and work of a man whom *The Economist* called 'the century's greatest champion of economic liberalism'.

The illustrations, including many previously unpublished photographs, draw heavily on family archives that were generously and patiently loaned by Dr Laurence and Mrs Esca Hayek, and by Miss Christine Hayek.

Thanks are also due to Dr Michael Walker of the Fraser Institute, who invited me to produce a slide and sound presentation on Hayek for the 1992 meeting of the Mont Pèlerin Society in Vancouver, which formed the foundation for the present work. An important stepping stone along the way was the Adam Smith Institute's video version of the presentation, called *Hayek: A Tribute*, which was produced with the encouragement and help of Linda Whetstone and the Atlas Economic Research Foundation (UK). The Lynde and Harry Bradley Foundation, and the Earhart Foundation, gave early support to my efforts to turn these audio-visual presentations into the present book.

Wherever possible, the events in Professor Hayek's life are described using his own words. *These appear in the text as italics.* The main sources, which are gratefully acknowledged, include *Hayek on Hayek*, edited by Stephen Kresge and Leif Wenar, a 1985 television interview between Hayek and John O'Sullivan, and Hayek's autobiographical lecture of 1983 entitled *The Rediscovery of Freedom*. Other sources are acknowledged in the notes.

Thanks are also due to Professor Donald Denman, without whose firm support the book would not have appeared; to Shirley Chapman at the LSE, who provided many useful suggestions; to the staff at the Saffron Walden public library, who located many sources; and to Sukie Hunter, who designed the layouts so creatively. Special thanks go to my wife, Heather, for her critical appraisal and extensive word-processing, and for tolerating the inevitable mountain of boxes of photographs and documents.

Finally, Dr Eamonn Butler, Director of the Adam Smith Institute, has been a most helpful editor and source of encouragement.

I hope this *Commemorative Album* will provide a unique insight, set against the background of the turbulent events of the 20th century, into the unfolding career and personal life of a scholar whose work and ideas have beneficially affected the lives of millions of people, even if most of them never knew his name.

John Raybould

Contents

1 Early years in Imperial Vienna, 1899–1914

Hayek's upbringing in an 'academic atmosphere'

Friedrich August von Hayek was born on 8 May 1899 into a gifted, academic family in Vienna, the vibrant and cosmopolitan capital of the powerful Austro-Hungarian Empire.

His father, Dr August von Hayek, was a medical doctor who became an honorary Professor of Botany at the University of Vienna and published notable works on the flora of the Austrian Alps. The young Friedrich was to develop an early fondness for those same mountains, which were to provide him with a lifetime of opportunities for walking, climbing, skiing and (above all), the solitude to think.

His mother, Felicitas (née von Juraschek), brought up her three sons in what he later described as an '*academic atmosphere*' in which the serious business of learning often seemed to take precedence over frivolous play. They met scholars in many disciplines who visited their parents' and grandparents' homes. All three Hayek brothers were destined to become professors.

Friedrich and his brothers grew up in the secure pre-war life at the heart of the Austro-Hungarian Empire. In 1908, Franz Joseph II celebrated the 60th year of his reign as Emperor of Austria with a grand reception, at which the music of the Strauss family reflected the gaiety and confidence of a society that was soon to vanish.

Fifty years later Hayek would write: '*To us that first decade of our century may seem a far away period of peace; and even in Central Europe the majority of people deluded themselves about the stability of their civilisation*'.[1]

Academic roots

Ever since the family was ennobled by Emperor Joseph I in the late eighteenth century, Hayek's ancestors almost invariably became civil servants, but they maintained a profound interest in the natural sciences too.[2]

Friedrich's grandfather, Gustav von Hayek, studied natural history and biology and eventually became a teacher at a *Gymnasium* (high school). Some of his systematic works on biology became fairly well known. This scientific interest was continued by Friedrich's two younger brothers (one an anatomist and the other a chemist), his daughter (an entomologist) and son (a medical microbiologist).

His other grandfather, Franz von Juraschek, had been a university Professor of Constitutional Law and later a top-ranking civil servant, and President of the Statistical Commission of Austria.

A wealthy family, the von Juraschecks lived in one of the most beautiful apartments in Vienna, across the Kärtnerstrasse from the Opera and facing the Ringstrasse. *'My grandparents' flat was a second home to me'* reminisced Hayek later, *'and the family present at … Sunday gatherings was large and would include a continuous range of ages from my grandparents to my youngest cousins.'*[3]

Franz von Juraschek (1849–1910).

Kärtnerstrasse, one of Vienna's principal streets, at the turn of the century.

'In this apartment in 14 Messenhausergasse at the corner of Landstrasse Hauptstrasse, on May 8 1899, Fritz Hayek saw the light of day.' Caption by his mother, Felicitas, under this photograph in the family album.

Hayek on his early upbringing

'My parents were exceedingly well suited to each other, and their married life seemed (not only to me) one of unclouded happiness .... My parents, though they had never formally left the ancestral Roman Catholic Church, held no religious beliefs ... all positive dogma was for them a superstition of the past'.[4]

'My father always hoped someday to be able to give up medicine altogether for a full university chair in botany, but that day never came, and the "Professor" was never more than the honorary title usually conferred on a Privatdozent *[lecturer] of several years standing. But while this unfulfilled ambition was a grief to him (and probably did much to make me regard a university chair as the most desirable of all positions I might attain), his scientific output was considerable, and in his particular field, plant geography (which today would be called ecology), he was highly respected by his fellows .... He died relatively early, in his fifty-seventh year from a severe blood poisoning [contracted on a botanical excursion]'.*[5]

Hayek with his mother, Felicitas von Juraschek, who lived until her 93rd year in 1967.

ABHANDLUNGEN
DER
K. K. ZOOL.-BOTAN. GESELLSCHAFT IN WIEN.
BAND IV, HEFT 2.

VORARBEITEN
ZU EINER
PFLANZENGEOGRAPHISCHEN KARTE
OESTERREICHS

IV.
DIE SANNTALER ALPEN
(STEINER ALPEN)

VON

Dr. AUGUST VON HAYEK
PRIVATDOZENT FÜR PFLANZENGEOGRAPHIE AN DER WIENER UNIVERSITÄT

MIT 14 ABBILDUNGEN UND EINER KARTE IN FARBENDRUCK

EINGEREICHT AM 5. OKTOBER 1906. – AUSGEGEBEN AM 24. SEPTEMBER 1907.

JENA
VERLAG VON GUSTAV FISCHER
1907

One of August von Hayek's published works on the flora of the Austrian

Hayek's father, Dr August von Hayek (1871–1928), was a Doctor of Medicine who turned to research and teaching, becoming an eminent botanist.

1902: Friedrich (left) and Heinz (right) with their mother, Felicitas.

1904: Friedrich with his baby brother Erich.

1903: Friedrich aged four.

All three Hayek brothers were destined to become professors: (left to right) Erich, Professor of Chemistry, University of Innsbruck; Heinz, Professor of Anatomy, University of Vienna; Friedrich, Professor of Economics at the Universities of London and Freiburg and Professor of Moral Philosophy at the University of Chicago.

'We probably had an ideal family life. Three meals together every day, talking about every subject under the sun, always left free by our parents to roam, to think, even to commit minor peccadilloes'.[6]

1908: left to right: Friedrich, Heinz, Erich.

1907: Hayek (centre back row) with his father, August, and Grandparents von Hayek in Gaishorn.

1906: Near Schladming in the Austrian Tyrol, left to right: Friedrich, Frau Dr Perlich, Heinz, Erich.

'My first scientific interest was, following my father, in botany.... I had much opportunity to help him, first as a collector and later as a photographer.... Systematic botany with its puzzle of the existence of clearly defined classes proved a useful education. But my interest gradually shifted from botany to palaeontology and the theory of evolution. I must have been about sixteen when I began to find Man more interesting, and for a time played with the idea of becoming a psychiatrist. Also public life and certain aspects of social organisation – such as education, the press, political parties – began to interest me, not so much as subjects for systematic study but from a desire to comprehend the world in which I was living'.[7]

1909: The nine-year-old Friedrich in a family group taken at his grandfather Juraschek's 60th birthday. Hayek is seated third left with his brother Erich. Hayek's father and mother are on the left, grandfather Juraschek is standing on the right. Hayek's brother Heinz is on the right between his two aunts, Gertrude and Greta.

1915: Hayek aged sixteen.

'I passed through a variety of schools, changing the Volkschule *[elementary school] once because of a move and* Gymnasium *[high school] twice because I ran into difficulties with my teachers, who were irritated by the combination of obvious abilities and laziness and lack of interest I showed. Except for biology, few of the school subjects interested me, and I consistently neglected my homework, counting on picking up enough during lessons to scrape through …'.*[8]

1916: Hayek (right) with his mother and brothers Erich and Heinz on their summer holiday.

The culture of ideas in Vienna

Hayek's cousin, Ludwig Wittgenstein, was destined to become a famous philosopher. Hayek would come to know him much later in Cambridge.

'I became probably one of the first readers of [Wittgenstein's] Tractatus *when it appeared in 1923 .... it made a great impression on me'.*[9]

The young Hayek also met the great liberal economist Eugen von Böhm-Bawerk, as a friend of his grandfather. Böhm-Bawerk was an economist in the tradition of Carl Menger and had been the Austro-Hungarian Empire's Minister of Finance in 1895, 1897 and 1900–1904. But all that was long before the young Friedrich even knew the meaning of the word 'economics'.

Ludwig Wittgenstein (1889–1951), Hayek's cousin.

'The beginning of my definite interest in economics I can clearly date back to a logic lesson in the seventh form of the Gymnasium, *late in 1916, when the master explained to us the threefold Aristotelian division of ethics into morals, politics and economics .... My reading then was mainly ephemeral pamphlets of contemporary politics, mostly of a socialist or semi-socialist character, from Karl Renner to Walter Rathenau, from the latter of whom I derived most of my first economic ideas'.*[10]

Eugen von Böhm-Bawerk (1851–1914), commemorated on the Austrian 100-schilling note.

2 World war and revolution, 1914–1918

Hayek experiences Europe in turmoil

On 28 June 1914, the assassination at Sarajevo of the heir to the Emperor of Austria, Archduke Ferdinand, triggered a chain of events that convulsed Europe. The armies of the Allied Powers of Great Britain, France, Serbia, Russia, Italy and later the United States were ranged against the armies of Germany and the Austro-Hungarian Empire. The First World War devastated Europe, leaving millions of young men dead or wounded and hundreds of towns destroyed or damaged.

As a young officer serving in an Austro-Hungarian field artillery battery on the Italian Front, Hayek saw war at first hand and watched the dramatic changes that were happening to the old world order.

In Eastern Europe, Russia was gripped by communism after the Revolution of 1917. Following the peace settlement after the Treaty of Versailles of 1919, Kaiser Wilhelm II fled from Germany and the Weimar Republic was created. In South-Central Europe, the Hapsburg Empire collapsed. New borders divided new nations. Czechoslovakia, Hungary and the republic of Austria were created and tried to build new political and economic relationships from the ruins.

Like so many young men who wanted to see a new world rise up from these ruins, the 19-year-old Hayek found himself powerfully attracted to what he called '*a mild Fabian socialism*'.

War shapes Hayek's political interest

During the war, Hayek served in the Austrian army, on the Italian front. In his words:

'I joined a field artillery regiment at Vienna in March 1917, and after a little over seven months' training was sent as a sergeant-major-officer-cadet (if I can thus translate the even longer title) to the Italian front where I served for a little over a year.... The most exciting moments were one abortive offensive in June 1918, the collapse of the Austro-Hungarian army in October 1918, and the two waves of retreat. Most of the later part of the period we were more concerned with hunger, disease, and the rumours of a mutiny of the Czechs than with serious fighting.'[1]

Hayek would later amuse audiences with his stories of army life – such as when he had to try to recapture a bucketful of eels which were meant for the breakfast of the troops, but which he had accidentally overturned in a dewy field.

The battle of Caporetto in October–November 1917 saw the Italian Second Army virtually wiped out by the Austro-German advance. However, a year later, Austria suffered a terminal blow delivered by the Italians after the Piave offensive.

'In the autumn of 1917, during the short leave from the army that I took in order to obtain my matura, *I actually got into trouble in the* Gymnasium *when I was found reading a socialist pamphlet during the divinity lesson. It was only sometime in 1917 or 1918, during a quiet period on the field on the Piave, that one of the slightly senior officers in my battery gave me the first systematic books on economics.... I think the decisive influence was really World War I, particularly the experience of serving in a multinational army, the Austro-Hungarian army. That's when I saw, more or less, the great empire collapse over the nationalist problem. I served in a battle in which eleven different languages were spoken. It's bound to draw your attention to the problems of political organisation. It was during the war service in Italy that I more or less decided to do economics.'*[2]

The young Hayek in his officer's uniform.

At the front in battle dress.

Europe moves towards socialism

As a reaction to the 'laissez faire' capitalism of the 19th century, radical alternatives developed. First the Ricardian socialists, then Marx and Engels and finally the Fabians advocated the restructuring of society. From these socialists stemmed the ideas that capitalism must inevitably fail, that workers were exploited and that revolution would empower the masses.

'I believe it was Lenin himself who introduced to Russia the famous phrase "who, whom?" – during the early years of Soviet rule the byword in which the people summed up the universal problem of a socialist society. Who plans whom, who directs and dominates whom, who assigns to other people their station in life, and who is to have his due allotted by others? These become necessarily the central issues to be decided solely by the supreme power.'[3]

The Russian Revolution, 1917: Lenin addressing a crowd in Moscow.

The work of Sydney and Beatrice Webb, co-founders of Fabian socialism in late 19th-century Britain and founders of the London School of Economics in 1895, made a considerable impression on the young Hayek.

'Originally I was, as every young intellectual, attracted by socialism of a mild sort. I probably was, when I began my study, because of what in England they would have called the Fabian kind of approach, convinced that there must be an intelligent solution of the many unsatisfactory events of this world.'[4]

Collectivism received a further impetus with the First World War, when the combatant countries created war economies based on central planning. The war was a dramatic test of the old European order, and was followed by revolution in many countries.

The Webbs: Beatrice (1858–1943) and Sydney (1859–1947).

3 At the University of Vienna, 1918–1926

Hayek's association with Ludwig von Mises

Friedrich Hayek entered the University of Vienna in November 1918. His student records show that he joined the Faculty of Arts, where he studied philosophy, law and economics. His teachers included leading members of the 'Austrian School' of economics, such as Professor Friedrich von Wieser, the successor to its founder Carl Menger. Hayek flourished under their guidance, abandoning his 'mild socialism' and discovering his interest in economic liberalism. By the age of 21 he already had his doctorate in Law; and two years later, in 1923, he was to gain his second doctorate in Political Economy.

Hayek's academic qualifications and enthusiasms were to prove invaluable to another of his mentors, Ludwig von Mises, who had written a major critique of socialism and regularly explored the principles of the market economy in his *Privatseminar* in Vienna. Mises, as well as being a teacher, was an Austrian civil servant running the temporary *Abrechnungsamt* (Office of Accounts) to carry out the financial provisions of the Treaty of St Germain in 1920, which broke up the Habsburg Empire after the First World War. He invited Hayek to join him as a legal consultant and their lifelong association and friendship began.

Hayek with fellow students at the University of Vienna, c. 1919.

After a short period of absence as a postgraduate research student at New York University in 1923–1924, Hayek returned to Vienna to resume his position with Mises at the *Abrechnungsamt*. He married in 1926 and a year later he helped Mises establish what was to become an influential research institute that would launch him on to the world stage as an economist.

'It was the political excitement connected with the First World War and the collapse of the Austro-Hungarian Empire which shifted my interests from a purely natural science background interest to a political interest. It was only at the very end that I decided to study law in order to specialise in economics. But it was the political excitement of the time which opened to me a new world.'[1]

Vienna in the period following the First World War was the intellectual centre of Europe, an enormously exciting place, bursting with ideas and energy.

'The University of Vienna, which I entered late in 1918 as a raw youth fresh from the war, and particularly the economics part of its law faculty, was an extraordinarily lively place. Though material conditions were most difficult and the political situation highly uncertain, this had at first little influence on the intellectual level preserved from pre-war days.... [T]he University of Vienna, which until the 1860s had not been particularly distinguished, then for a period of sixty or seventy years became one of the intellectually most creative anywhere and produced distinct internationally known schools of thought in a great variety of fields: philosophy and psychology, law and economics, anthropology and linguistics, to name only those closest to our interests.'[2]

Hayek's student record book shows the academic career of a very bright student who apparently attended every lecture that was going. Such books were used in Vienna so that once the professors in question had signed them, the students then had to pay for the lectures!

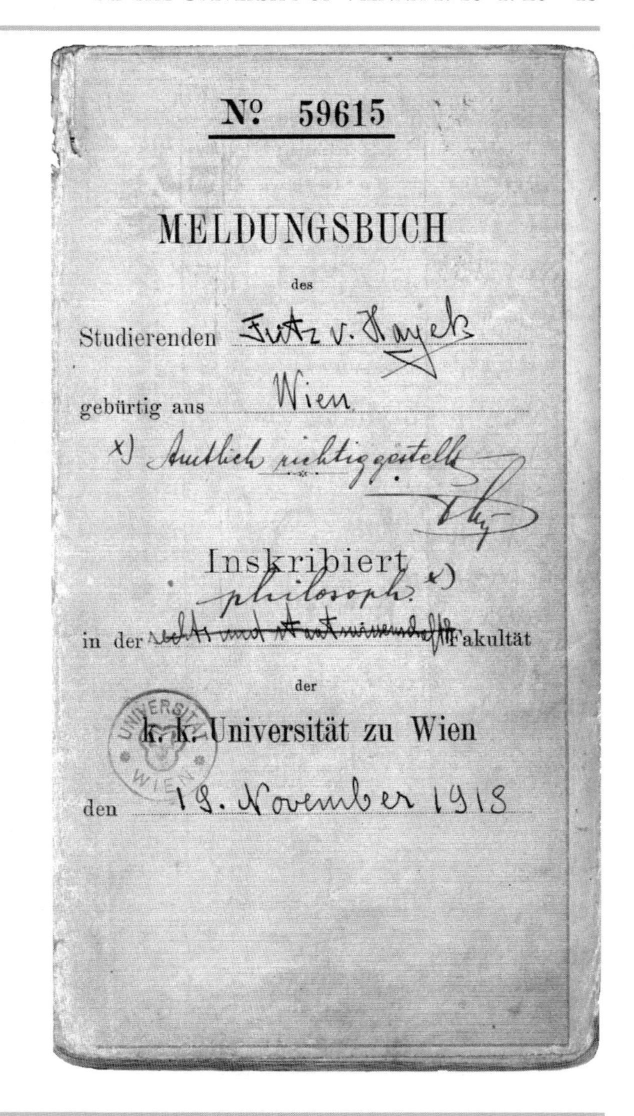
№ 59615

MELDUNGSBUCH

des

Studierenden Fritz v. Hayek

gebürtig aus Wien

Inskribiert

in der ... Fakultät

der

k. k. Universität zu Wien

den 18. November 1918

Hayek's original student record book, never before reproduced in print, shows that he joined the Faculty of Arts, where he studied Philosophy, Law and Economics.

Mises and others signed Hayek's student record book for the May–July term 1922.

Hayek studies under leading Austrian School economists

Carl Menger (1840-1921) the author of *Grundsätze der Volkswirtschaftslehre* (Foundations of Economics) and founder of the 'Austrian School' of economics.

Hayek's tutors included leading lights of what was known as the 'Austrian School' of economics, founded by Carl Menger. Hayek flourished under their guidance, discovering his interest in economic liberalism which was to stay with him all his life.

Among the economics teachers who signed Hayek's student record book were Friedrich von Wieser (1851–1926), Professor of Economics at the University of Vienna, and Othmar Spann (1878–1950), philosopher and founder of 'universalist economics' at the University of Vienna.

'Wieser, the last living link with the great past, seemed to most of us at first a somewhat aloof and unapproachable grand-seigneur. *He had only just returned to the University after serving as Minister of Commerce in one of the last imperial governments. He lectured on the lines of his* Social Economics... *the only systematic treatise on economic theory which the Austrian School had produced and which he seemed to know more or less by heart.'*[3]

'Spann... had some helpful things to say on the logic of the means–ends relationship but soon moved into regions of philosophy which to most of us seemed to have little to do with economics. But his little textbook on the history of economics, reputedly modelled on Menger's lectures on the subject, was for most of us the first introduction to this field.'[4]

'Though a new degree in the political and economic sciences had just been created, most of us were still working for the law degree in which economics was only a small part and any professional competence we had largely to acquire by our own reading and from the teaching of men for whom this was a part-time labour of love. The most important of them was of course Ludwig von Mises, but I myself came to know him well only comparatively late.'[5]

In the winter of 1919–1920 the University of Vienna was closed because of the lack of heating fuel. Hayek became politically aware just as the new Austria of those postwar days found itself in the grip of severe inflation and even starvation.[6]

In 1921 Hayek obtained his PhD in Law and became 'Dr Jur'. Two years later he obtained his PhD in Political Economy and became 'Dr Rer Pol'.

Ludwig von Mises

Ludwig von Mises received a PhD in Law from the University of Vienna in 1906. He was an Austrian civil servant and economic philosopher on whom had fallen Menger's mantle as leader of the Austrian School.

Mises had the courtesy title of *Privatdozent* at the University and ran what was known as his *Privatseminar*, in which gifted young academics met and discussed important problems of economics, sociology and philosophy. Hayek attended regularly and said later:

'Mises's Privatseminar *was really entirely outside the university. These were fortnightly informal meetings held in the evening at Mises's office at the Chamber of Commerce and invariably continued far into the night at some coffee house… from about 1924 to 1931, assisted by the circumstance that Mises had got Haberler and myself jobs in the same building… the meetings held there were at least as much the centre of economic discussion in Vienna as the University.'*[7]

Hayek's obvious intellectual abilities stood out among his contemporaries. He struck up what was to become a lifelong friendship with Mises, and it was through Mises that he began to lose the mild socialism that coloured his early political views.

'It was as a mild socialist that I decided I must study economics. I was very soon cured of this belief that socialism was the solution because I came after three years under the direct influence of Ludwig von Mises who had then [1922] published his great book Socialism, *demonstrating that the socialist solution was impossible in a technical sense.'*[8]

Ludwig von Mises (1881–1973).

Hayek and Mises enjoy a joke many years later.

Hayek began to realise that socialism promised more than it could ever deliver since it simply assumes that all the knowledge required to run a society can be collected and processed by some single authority. It overlooks the fact that modern society is based on the utilisation of widely dispersed knowledge, which cannot possibly be collected and collated by a central command structure.

'When Socialism *first appeared in 1922, its impact was profound. It gradually but fundamentally altered the outlook of many of the young idealists returning to their university studies after World War One. I know, for I was one of them.... Socialism promised to fulfil our hopes for a more rational, more just world. And then came this book. Our hopes were dashed.* Socialism *told us that we had been looking for improvements in the wrong direction.'*

Continuing, in his Foreword to the 1981 edition of the book, Hayek wrote: 'Socialism *shocked our generation and only slowly and painfully did we become persuaded of its central thesis.'*[9]

On this 21st birthday photograph his mother proudly wrote: 'Friedrich already has his Doctorate in Law'.

Hayek works for Mises at the *Abrechnungsamt*, 1921–1926

By 1921 Ludwig von Mises was in charge of a temporary Austrian government office called the *Abrechnungsamt* (Office of Accounts), established to carry out the terms of the Treaty of St Germain as they affected the new country of Austria, which had been created following the dismemberment of the Habsburg Empire after the war.

Hayek's qualifications were to prove invaluable to Mises when, in 1921, he took his first job, as a legal consultant to Mises in the *Abrechnungsamt*. Hayek's knowledge of French and Italian, and later of English, together with his knowledge of law and economics, qualified him for what was a comparatively well-paid job. This association was to have a profound effect on Hayek's whole life and career. Hayek, many years later, described Mises as '*one of the best educated and informed men I had ever known and, what was most important at the time of great inflation, as the only man who really understood what was happening.*'[10]

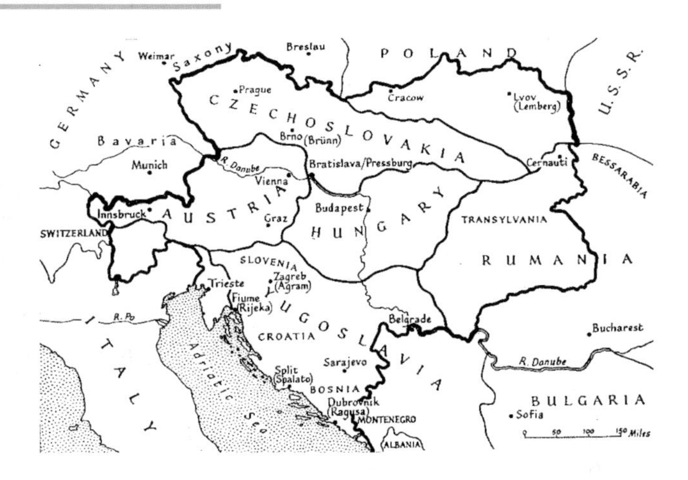

New borders and new countries – Czechoslovakia, Rumania, Hungary, Yugoslavia and the Republic of Austria – meant that Vienna was no longer the heart of an empire.

ABRECHNUNGSAMT, WIEN.

Legitimationskarte Nr. 63

für Dr. Fritz H a y e k,

Sekretär d. Abrechnungsamtes

wohnhaft: Wien, V. Margarethenstrasse 82

Wien, am 6. März 1926.

Das Abrechnungsamt.

Eigenhändige Unterschrift:

Hayek's pass to the *Abrechnungsamt*.

Hayek (left) with colleagues in the *Abrechnungsamt*.

Hayek's New York interlude, 1923–1924

Given leave of absence from the Austrian Federal Service in 1923, Hayek had decided that a visit to the United States was essential for an aspiring young economist and scraped together the funds for the journey. He even secured a half-promise of a job over there. Wieser then asked the Viennese banker and economist Joseph Schumpeter to give Hayek letters of introduction to his friends in the States… all the great American economists.

'[The letters] did prove true "open sesames" – as probably the first Central European economist to visit the States after the war I was received and treated much beyond my deserts by John Bates Clark at New York, T Carver at Harvard, Irving Fisher at Yale, and Jacob Hollander at Johns Hopkins.

'…When my expectations of a job failed and my small funds were used up, I never had to start on the job as a dishwasher in a Sixth Avenue restaurant which I had already taken but was found an assistantship with Jeremiah Jenks [the eminent international monetary economist] *at New York University… which enabled me to devote my time to more intellectual matters.'*[11]

'One of the first conclusions at which I remember I had arrived towards the end of 1923 was that stabilisation of national price levels and stabilisation of foreign exchange were conflicting aims. But before I could anywhere submit for publication the short article I had written on the subject, I found that Keynes had just stated the same contention in his Tract on Monetary Reform *(1923). Lest anybody thinks that this disappointment in my hope of having made an original discovery is responsible for my later persistent opposition to Keynes, I should add that Keynes was then, and remained for a good deal longer, one of my heroes and that I greatly admired this particular work of his.'*[12]

The main branch, New York Public Library, where Hayek did most of his work.

In Central Park, New York, 17 December 1923 (with his trademark pipe!).

The 1920s hyperinflation in Europe

The great inflation in Germany after the First World War, which spilled over into Austria too, was one of the most remarkable events of the century. Currency in circulation in Germany rose from 6 billion marks in 1913 to 92,000,000 billion in November 1923. A pair of shoes costing 12 marks in 1913 was selling for 32 trillion marks in November ten years later. Hayek himself recalled that he had been given 200 pay rises in eight months merely to keep up with prices that doubled every day.[13]

During this disastrous inflation (which Keynes claimed resulted from the Allies' punitive war reparations), Hayek wrote a letter (right) to *The New York Times* on monetary instability and Germany's finances. It was an early example of what became a regular flow of 'Letters to the Editor' that Hayek was to write on many aspects of political economy over the next 65 years, and shows the origins of his life-long and deep-rooted fear of rampant inflation and the social upheaval it causes. It also shows his remarkable command of written English (he was just 24).

The young Hayek put his time in the United States to good use professionally. But Europe beckoned, so he abandoned his research in America. He returned quickly to Vienna in the summer of 1924 to take up his former position as a legal consultant in Mises's *Abrechnungsamt* and, more importantly, to draw on his experience in America to set up a business-cycle research institute with Mises.

Germany's Finances.

• • • It is a matter of common knowledge that Germany's middle classes, including the small merchant and manufacturer, have lost almost everything. Only the big industries, which are organized there as in other countries as joint stock companies, might be supposed to have enriched themselves. Their earning capacity, however, is naturally represented in the market price of their stock, which is not affected directly by the value of the mark, but by the expectation of future profits. Though the average price of the shares, quoted on the Berlin Stock Exchange, as expressed in marks, rose with every fall of the mark, their gold value has fallen continuously since 1918. While the average gold value of these shares, as represented by an index number prepared by the Wirtschaft und Statistik, in 1918, was still 91 per cent.

of their value in 1913, it fell to 28 per cent. in 1919; 14 per cent. in 1920; rose during the stabilization of the mark in 1921 again to 18 per cent., and fell to 9 per cent. in 1922, and to 5.24 per cent. in January, 1923. Since then the mark has fallen to less than 1-500 of its value, so it is easy to imagine how far the impoverishment of Germany may have proceeded in the meantime.

F. A. HAYEK.

New York, Aug. 13, 1923.

Hayek (with beard) in a deck chair on the *Amsterdam* returning to Vienna after studying in New York.

Marriage and family life in Vienna

In 1926, Hayek married Helene ('Hella') von Fritsch. They were to have two children, Christine (born in 1929) and Laurence (born in 1934).

Baby Christine with her mother.

Hayek in the Vienna woods, about 1926.

Christine with her father.

4 The rising Austrian economist, 1927–1931

Hayek's work on business cycles makes his reputation

In 1927, after Hayek had returned from America, he and Mises founded the Austrian Institute for Business Cycle Research in Vienna, with Hayek as its first Director. Its Board of Trustees included many members of the Austrian School of economics.

During the 1920s Hayek increasingly entered the international economics arena and started coming into contact with leading economists. In 1928 he was invited to attend the London Conference on Economic Statistics, where he met for the first time John Maynard Keynes, already celebrated as an influential British economist. The two were to become lifelong friends as well as tenacious intellectual adversaries until Keynes's untimely death at the peak of his influence in 1946.

In 1929, in the midst of the Wall Street crash, the 30-year-old Hayek, as well as running the Austrian Institute for Business Cycle Research, took on the challenge of his first academic post, when he was appointed a *Privatdozent* in Economics and Statistics at the University of Vienna, a position he held until 1931.

Hayek, through his scholarly association with Mises and following in the tradition of Menger, von Wieser and the great Böhm-Bawerk, was rapidly becoming a leading figure of the Austrian School of economics. He began writing and publishing original and acclaimed works on economic theory, starting in 1929 with *Geldtheorie und Konjunkturtheorie* (*Monetary Theory and the Trade Cycle*).

When his 1929 inaugural lecture at Vienna came to the attention of Lionel Robbins at the London School of Economics, there were to be dramatic consequences for Hayek's career.

Hayek with a fellow participant at a conference in Rotterdam, 1930.

Mises and Hayek establish an economic research institute in Vienna

In 1927 Hayek and Mises started the *Österreichisches Institut für Konjunkturforschung* (Austrian Institute for Business Cycle Research). Hayek became the founding Director.

'Great fascination... was exercised at the time by the attempts at economic forecasting, particularly the economic barometers of the Harvard Economic Service, and, however questionable it all appears in retrospect, acquaintance with them and the whole technique of dealing with economic time series was the greatest practical advantage we returnees from the United States derived.... [I]t did produce the solid advantage that it forced us to familiarise ourselves with the modern techniques of economic statistics which were still practically unknown in Europe.

Hayek at his desk at the Institute in Vienna.

ÖSTERREICHISCHES
INSTITUT FÜR KONJUNKTURFORSCHUNG.

KURATORIUM:

Richard REISCH
Vorsitzender.

Walter BREISKY
Stellv. Vorsitzender.

Ludwig MISES
Stellv. Vorsitzender.

Franz BARTSCH, Wilhelm BERLINER, Eduard BREISACH, Rudolf BRICHTA, Guido BROMEISSL, René DELANNOY, Franz DOMES, Theodor DORAZIL, Hugo FORCHER, Karl FORCHHEIMER, Ernst GEIRINGER, Friedrich GLASER, Viktor GRAETZ, Richard KATZIANTSCHITSCH, Benedikt KAUTSKY, Richard KERSCHAGL, Alfred KOROMPAY, Otto KORWIK, Ludwig KRANZ, Helene LIESER, Edmund MARENZELLER, Hans MAYER, Edmund PALLA, Emil PERELS, Karl PRIBRAM, Alexander REICHMANN, Robert RICHTER, Richard RIEDL, Rudolf RIEMER, Hans RIZZI, Emil ROTHE, Walter SCHIFF, Karl SCHLESINGER, Hans SCHMIDT, Richard SCHUELLER, Friedrich SCHUSTER, Max SOKAL, Siegfried STRAKOSCH-FELDRINGEN, Ernst STREERUWITZ, Richard STRIGL, Ludwig STROBL, Wilhelm TAUCHER, Friedrich TILGNER, Josef VINZL, Wilhelm WINKLER, Otto WITTSCHIEBEN, Gustav WEISS-WELLENSTEIN, Heinrich WILDNER.

LEITER DES INSTITUTES:
Friedrich A. HAYEK.

'There can be no doubt that it was this experience on my American visit which turned me... to the problems of the relations between monetary theory and the trade cycle.'[1]

Mises's faith in Hayek's abilities as an economist, statistician and administrator was well placed. Years later, Margit von Mises said her husband always 'met every new student encouraged, hopeful that one of them might develop into a second Hayek. If he saw a tiny spark, he hoped for a flame.'[2]

Eminent economists of the Austrian School feature on the Institute's board.

The London Conference on Economic Statistics, 1928

Hayek's international links and reputation began to develop and to grow in the late 1920s when he started coming into contact with many of the great economists of the age. He met John Maynard Keynes for the first time at a conference in London. Keynes was at the height of his fame as a British economist while Hayek was a relatively unknown young Austrian economist.

'I had been the first to organise a meeting of Konjunkturinstituten *in Vienna, and then London repeated this and of course invited me. Keynes was a member of the board. We at once had a conflict, very friendly, about the rate of interest. Inevitably. He did, in his usual manner, try to go like a steamroller over the young man. But the moment – I must grant him this – the moment I stood up with serious arguments, he took me seriously, and ever since respected me.'*[3]

Keynes, with his wife, the famous ballerina Lydia Lopokova, in 1928.

Hayek and Keynes were to become personal friends, but they remained intellectual adversaries for nearly 20 years, until Keynes died at the comparatively early age of 63 in 1946 (Hayek was to survive him by nearly 50 years!).

The 1928 conference in London, on the subject of economic statistics, was attended by leading British economists, including Keynes, as well as by economists from Berlin, Budapest, Vienna and Harvard. Also present was Sir William Beveridge, at that time the Director of the London School of Economics. Perhaps this meeting had some bearing on the invitation to join the LSE that Hayek received three years later. Hayek said that he had already found *'uncommonly stimulating'* the writings of Edwin Cannan, Professor of Economics at the LSE until 1926.

VERTRAULICH

Protokoll
der
Londoner Konferenz
der
Konjunkturforschungsinstitute
Juni 1928

Gedruckt für die Mitglieder des Kuratoriums des Österreichischen Institutes für Konjunkturforschung
Druck: Buch- und Kunstdruckerei »Steyrermühl«, Wien VI

'Protokoll' ('Proceedings') of the 1928 Conference translated into German by Hayek and published by his Institute in Vienna.

William Beveridge, Director of the LSE and future architect of the UK welfare state.

Hayek's iconoclastic book on trade cycles

In the late 1920s Hayek began publishing works on economic theory, the first of a prodigious output throughout his long academic career.

He conducted a public and private debate with Keynes on the importance of money for the next 20 years. It was a subject that Hayek had definite views about, given his personal experience of hyperinflation in Austria.

Hayek became an acknowledged expert on monetary theory with the 1929 German language publication, by the Institute in Vienna, of his first book *Geldtheorie und Konjunkturtheorie*, published in English in 1933 as *Monetary Theory and the Trade Cycle*.

Written on the eve of the Wall Street crash, the book perceptively examined the role of money and the banks in causing economic fluctuations. It rejected the then current view that money and the availability of credit did not affect the structure of production, and it showed how a sudden injection of credit into the economy can cause changes in the relative prices between goods and lead to an over-investment that cannot be maintained.

The first six publications produced by Hayek's Institute included two by Hayek himself (*Geldtheorie* and later *Preise und Produktion*) and two by Fritz Machlup. Machlup (1902–1983) had been an original member of Mises's *Privatseminar* and went on to have a distinguished teaching career in the United States at the Universities of Buffalo and Princeton. Oskar Morgenstern (1902–1977), also a *Privatseminar* member and the author of another publication, followed Hayek as Director of the Institute in Vienna in 1931, moving on to become a Professor at Princeton in 1938 when the Nazis invaded Austria.

Hayek with his Institute's second author, Fritz Machlup, pictured at Beloit College around 1960.

Lecturer at the University of Vienna

In 1929 the 30-year-old Hayek, in addition to working for Mises at their new Austrian Institute and publishing his first book, took on the challenge of his first academic post. Appointed a *Privatdozent* in Economics and Statistics at the University of Vienna, he held the post until 1931.

Because of his growing intellectual association with Mises, he was now a thoroughgoing Austrian School economist, teaching in the tradition of Wieser, Menger and Böhm-Bawerk. His long teaching career was about to begin.

Hayek later acknowledged that it was luck that now brought about a major change in his life, when his inaugural lecture at the University of Vienna came to the attention of Lionel Robbins, the young, new head of the Department of Economics at the London School of Economics.

5 At the London School of Economics, 1931–1950

Hayek, Robbins and the long debate with Keynes

In 1929 the young Lionel Robbins became head of the Department of Economics at the London School of Economics. He was familiar with the Austrian School and particularly with the work of Friedrich Hayek.

Robbins recognised Hayek as able to provide a scholarly counterbalance to the theories of John Maynard Keynes and the others at the University of Cambridge, so in 1931 he invited Hayek to lecture at the LSE. A year later, having made a profound impression on everyone, Hayek was appointed Tooke Professor of Economic Science and Statistics at the University of London. Thus began his long and influential teaching and writing career at the London School of Economics.

Robbins encouraged Hayek to publish his original 'sensational' LSE lectures in 1931 as *Prices and Production*. Throughout the 1930s Hayek was to produce a stream of academic books, including *Profits, Interest and Investment*, *The Pure Theory of Capital* and (as editor and contributor) *Collectivist Economic Planning* and the works of Carl Menger.

In the 1930s Hayek chose to enter the wider debate on state planning, and joined in the academic debate on monetary, capital and business cycle theories that raged during the Great Depression. Hayek and Robbins became major figures in the celebrated London–Cambridge controversies with John Maynard Keynes, whose prolific output included the *Treatise on Money* and his highly influential *General Theory* in 1936.

The 1930s saw Friedrich Hayek settling into his new life in London with his wife Hella and his two young children, Christine and Laurence. These days in London were interspersed with the equal contentment of scholarly retreats to his beloved Austrian Alps.

In 1938 Hayek became a naturalized British subject. The coming of war would change the direction of his career yet again.

At the London School of Economics

Lionel Robbins (1898–1984),
Professor at the LSE 1929–1961.

THE PURCHASING POWER OF THE CONSUMER AND THE DEPRESSION.

The lot of the economist to-day is a hard one.
It is an unfortunate lot to-day to be an economist. Everybody else may
feel justified in taking a rather optimistic view at least of the somewhat
more remote future. Even if the immediate outlook is as gloomy as at present
the rapid progress of knowledge in the technical sciences seems to secure
a brilliant future to mankind and a speedy overcomming of our present economic troubles. To the economist the situation of his own branch of knowledge
gives no reason for such confidence. Not that progress has here been slower
than in other sciences. But it lacks that application to the problems of
actual life which is essential if our knowledge is to be ...
in other lines of knowledge the practitio...
treats are trained at least in its fundame...
such that the men who should apply economi...

This, if I remember rightly
was a lecture I gave at Cambridge
in January 1931 before going
to London to give the lectures
on Prices and Production
F.A.H. 2.X.64

In the 1930s, the Economics Department of the London School of Economics developed a substantial reputation under the vigorous leadership of Lionel (later Lord) Robbins, who was appointed to the Chair in 1929, aged only 30 – at the time the youngest professor of economics in England.

Robbins was to head the Department until 1961. During his tenure, particularly in the inter-war years, the Department acted as a magnet for bright students from many parts of the world and attracted eminent scholars from across Europe. One of these scholars was Friedrich Hayek, whom Robbins invited to deliver a set of lectures on economics at the LSE in 1931.

An intriguing chain of events had brought Hayek to London. His initial lectures as *Privatdozent* in Vienna examined critically the under-consumption theory that was gripping the contemporary economics debate in England. Robbins could read German and was sympathetic to the approach of the Austrian School. So Hayek was called in to help the intellectual fight against Keynes.

The manner of Hayek's actual arrival in England was brave, if not a little foolhardy. Prior to lecturing at the LSE, he went to Cambridge to deliver a one-lecture version of his theory on 'The Purchasing Power of the Consumer and the Depression' to the members of the Marshall Society. Here in the very heart of Keynes's territory, he told his audience that Keynes was wrong, that the slump was due to over-investment and that the cure lay in increasing saving. Faced by the younger generation of Cambridge economists in the so-called 'Cambridge Circus', who lionised Keynes (including Richard Kahn and the redoubtable Joan Robinson), it was no wonder that 'Hayek's exposition was greeted with complete silence.'[1]

Hayek delivering a lecture at the LSE (a previously unpublished photograph).

Arriving at the LSE in January 1931 to deliver his lectures, Hayek greatly impressed Robbins, as he had impressed Mises earlier: 'I can still see the door of my room opening to admit the tall, powerful, reserved figure which announced itself quietly and firmly as "Hayek",' noted Robbins in his autobiography.

'In the event the lectures were a sensation,' continued Robbins, 'partly for their revelation of an aspect of classical monetary theory which for many years had been forgotten, partly for the development of models of an elementary structure of a capitalistic economy which purported to show the influence on production and relative prices of changes in the proportions of expenditure allotted to consumption and investment respectively. The lectures were at once difficult and exciting… they conveyed such an impression of learning and analytical invention that when, greatly to my surprise, Beveridge [the Director of the LSE] asked if we would care to invite the lecturer to join us permanently… there was a unanimous vote in favour. To the delight of all concerned, Hayek was willing to accept.'[2]

In 1931, Hayek was therefore appointed as Visiting Professor, for one year, to the revived Tooke Chair as Professor of Economic Science and Statistics at the University of London. His position became permanent a year later and Hayek remained at the LSE until he went to the University of Chicago in 1950.

His Inaugural Lecture, 'The Trend of Economic Thinking', outlined Hayek's own view of the role of the economist in intellectual life, and was an early public statement of why he disagreed with socialism. *'Most of the planners'*, he commented, *'do not yet realise that they are socialists.'*

VER, SUNDAY, JANUA

ERSITIES.

LONDON.

THE TOOKE CHAIR.

(By Our Own Correspondent.)

Dr. F. A. von Hayek, of Vienna, has been appointed Tooke Professor of Economic Science and Statistics at the School of Economics. The appointment is for one year. He has been given a year's leave by the "Osterreichisches Institut für Konjunkturforschung," of which he is director. He is thirty-one years old, was educated at Vienna University (Dr. Jur., 1921, Dr. Sci. Pol., 1923) and at Zürich. He has also studied at New York University and the National Bureau of Economic Research. He served for one year (1924-5) in the Austrian Government Service and has had teaching experience at Vienna University as "Privatdozent" and at the "Handelschochschule." He gave advanced lectures in economics at the London School of Economics early last year.

Hayek's first book published in England

Hayek's four lectures at the LSE were so relevant to the debate on the causes of the severe deflationary slump that was currently plaguing Western economies that, with Robbins's encouragement, they were quickly published in 1931 in London by Routledge as *Prices and Production*.

PRICES AND PRODUCTION

BY

FRIEDRICH A. HAYEK

Lecturer in Economics in the University of Vienna
Director of the Austrian Institut für Konjunkturforschung

WITH A FOREWORD BY

LIONEL ROBBINS

Professor of Economics in the University of London

LONDON
GEORGE ROUTLEDGE & SONS, LTD.
BROADWAY HOUSE: 68-74 CARTER LANE, E.C.
1931

The lectures were also published by Hayek's Institute in Vienna (and later in the US in 1932, Japan in 1934, Taipei in 1966, and France in 1975).

With the Wall Street crash a very recent event, Professor Robbins said in his Foreword: 'The pure theory of economic equilibrium, the great achievement of nineteenth-century economics, provides no explanation of trade depression. It explains the tendencies conducive to stability in the economic system. ...But it does not explain the occurrence of periodic disequilibrium....

'Professor Mises and Dr Hayek have advanced theories which, though they fall into the general category of monetary explanations, yet seem altogether free from those deficiencies which have marked monetary explanations in general....

'Dr Hayek... does not claim to provide a cut and dried cure for all the evils of the monetary system... [but] I am bound to say that it seems to me to fit certain facts of the American slump better than any other explanation I know. And I cannot think that it is altogether an accident that the Austrian *Institut für Konjunkturforschung*, of which Dr Hayek is Director, was one of the very few bodies of its kind which, in the spring of 1929, predicted a setback in America, with injurious repercussions on European conditions'.

Der Leiter des österreichischen Instituts für Konjunkturforschung, Dozent Dr. Friedrich Hayek, erhielt für das Studienjahr 1931/32 eine Berufung an die London School of Economie and Political Science.

Press announcements of Hayek's appointment at the LSE in 1931 appeared in Vienna also.

The influence of the LSE

Sir William (later Lord) Beveridge (1879–1963) was Director of the London School of Economics between 1919 and 1937. Under Beveridge, said Hayek, '*the LSE became in the 1930s perhaps the most lively centre of economic discussion… [and there] developed a strong contrast between the somewhat insular, purely Marshallian tradition of Cambridge and Oxford and the truly international synthesis of London.*'[3]

Hayek undoubtedly played a part in making it so. Robbins later described him as 'a major stimulus to thought' at the LSE 'for nearly 20 years'.

'His own thought had the dual qualities of depth and of great originality. Both in fundamental analysis and in the theory of policy, he lived at the frontiers of speculation…. But his work was not only important, it was also very stimulating: whether you agreed with him or not, you could not talk to Hayek without being induced to think for yourself. Contrary to popular belief, as a teacher Hayek was no proselytizer. He had strong convictions himself. But in discussion his focus was always directed not to persuade but rather to pursue implications.'

Robbins continued: 'For me personally, the association was an especially happy one…. We shared many intellectual and cultural interests outside our professional specialisation; and although we frequently differed in matters of practical judgement, we had a common devotion to the ideals of a free society and a common apprehension of the contemporary dangers thereto.'[4]

Hayek's 'official' portrait taken in London in 1931.

Hayek at the door of the LSE during a visit in the 1980s.

Ronald Coase, who taught at the LSE from 1935–1951, before going on to the University of Chicago, where he won the Nobel Prize in Economics in 1991, concurred with Robbins regarding the importance of Hayek's contribution in encouraging intellectual rigour at the LSE: 'Unassertive, Hayek nonetheless exerted considerable influence through his profound knowledge of economic theory, the example of his own high standards of scholarship and the power of his ideas.'[5]

Ronald Coase, an LSE student from 1929–1932.

These were vibrant times for the LSE and Hayek: '*In some ways it is these years in London before the war which in retrospect seem to me the intellectually most active and satisfying of my life. I certainly never again could arouse the same passionate interest in the technicalities of theoretical economics or profit in the same way from discussion with first-class minds with similar interests. Especially the seminar – which was really conducted by Robbins but for which I nominally shared responsibility with him – taught me more economics than anything else.*'[6]

The LSE's famous entrance in Houghton Street, London.

Just as Mises had his *Privatseminar* in Vienna, so Robbins and Hayek led their own weekly seminar in London. 'We endeavoured to keep under review the main developments in our sector of the frontiers of knowledge and to make our own contributions thereto,' wrote Robbins. 'It was all very exciting.'[7]

The seminar was addressed not only by visiting economists from British universities but also by many visiting economists from abroad, including Gottfried Haberler and Fritz Machlup, both from Vienna, Jacob Viner from Princeton and Frank Knight from Chicago.

J K Galbraith described Hayek's seminar (which he attended in 1937 and 1938) as 'possibly the most aggressively vocal gathering in all the history of economic instruction'. But, he added, 'Hayek was one of the gentlest in manner, most scholarly and generally most agreeable men I have known.'[8]

London versus Cambridge

Vigorous, sometimes bellicose, debate between the economists at the London School of Economics and at the University of Cambridge went back to the early years of the LSE in the 1900s, when Edwin Cannan in London and Alfred Marshall in Cambridge argued about the very foundations of the subject.

Although in the 1920s the economists in London and Cambridge collaborated regularly on routine activities, such as the London and Cambridge Economic Service, and acted as external assessors in each other's final examinations, by the 1930s a second dispute had erupted.

Ralf (later Lord) Dahrendorf, who was Director of the LSE between 1974 and 1984, saw the argument revolving around 'Lionel Robbins (as well as Friedrich Hayek) and John Maynard Keynes.... [It] initially had to do with deflation versus demand management as a remedy for the depression, but soon also involved different notions of economics and notably what came to be called macroeconomics. LSE lost on both counts, though it could be argued that each time it prevailed in the long run... and, more importantly,... it was the livelier, more pluralistic and open centre for teaching and research throughout the four decades of LSE–Cambridge "warfare".'[9]

Edwin Cannan (1861–1935), Lecturer and Professor of Economics at the LSE 1895–1926 and editor of a famous edition of Adam Smith's *Wealth of Nations.*

Alfred Marshall (1842–1924), Professor of Political Economy at Cambridge 1885–1908. The leading economist of his times, he was author of the influential *Principles of Economics.*

The hustle and bustle of traffic by the Bank of England in the centre of London in the 1930s contrasts with the age-old traditions of King's College Cambridge.

The long debate with Keynes, 1928–1946

Thirty years on, Hayek reflected: *'When I look back to the early 1930s, they appear to me much the most exciting period in the development of economic theory during this century. This is probably a highly subjective impression, determined both by my age at that time and the particular circumstances in which I was placed. Yet even when I try hard to look at the period as objectively as I can, the years between about 1931, when I went to London, and say 1936 or 1937, seem to me to mark a high point and the end of one period in the history of economic theory and the beginning of a new and very different one. And I will add at once that I am not at all sure that the change in approach which took place at the end of that period was all a gain and that we may not some day have to take up where we left off then.'*[10]

Hayek remained the most ardent critic of the economic planning and interventionism that was inherent in the approach of Keynes and his followers. Their long intellectual 'battle' really began with the publication in 1930 of Keynes's *A Treatise on Money.*

Hayek's intellectual style was one of great rigour. This put him at considerable disadvantage against the quick-witted and ebullient Keynes, 16 years his senior, editor of the *Economic Journal* since 1911, secretary of the Royal Economic Society since 1913 and author of the 1919 best-seller, *The Economic Consequences of the Peace.*

After meeting Keynes for the first time in 1928, Hayek remained good friends with him thereafter. They had many interests in common, although Hayek later observed that *'we rarely could agree on economics'*.

Keynes was adept at shifting his ground when the economic circumstances of the time demanded new policy prescriptions. Hayek, along with a number of Keynes's economist friends at Cambridge, had very severely reviewed the *Treatise.* Hayek had been dismayed that, just when he thought he had finally demolished the relation between aggregate demand and employment, it had re-emerged in Keynes's book. After spending a great deal of time and effort on his critique he said he was exasperated to be told by Keynes, *'Oh, never mind; I no longer believe all that.'*

Keynes, meanwhile, condescendingly described *Prices and Production* as 'one of the most frightful muddles I have ever read, with scarcely a sound proposition in it'. However, he added, 'Yet it remains a book of some interest which is likely to leave its mark on the mind of the reader.'[11]

In 1936, Keynes published a controversial book that would completely change the economic and political landscape of the world for the next 40 years: his *General Theory of Employment, Interest and Money.*

Largely because he suspected that Keynes would change his mind yet again, Hayek did not attempt a systematic refutation of this, Keynes's most influential work.

'I feared that before I had completed my analysis he would again have changed his mind. Though he had called it a "general" theory, it was to me too obviously another tract for the times, conditioned by what he thought were the momentary needs of policy.'

It was a mistake for which Hayek bitterly blamed himself in later years. And yet, *'more than any other single work it [the* General Theory*] decisively furthered the ascendancy of macroeconomics and the temporary decline of microeconomic theory.'*[12]

'I wish I knew why the General Theory *had such an enormous influence. I was puzzled at the time and, in fact, I did not believe it would succeed.'*[13]

COLLECTIVIST ECONOMIC PLANNING

Critical Studies on the Possibilities of Socialism by N. G. PIERSON, LUDWIG von MISES, GEORG HALM, and ENRICO BARONE

Edited, with an Introduction and a Concluding Essay by

F. A. von HAYEK

LONDON
GEORGE ROUTLEDGE & SONS, LTD.
BROADWAY HOUSE, 68–74 CARTER LANE, E.C.
1938

Though really an academic economist, Hayek entered the wider political debate over the merits of state planning in 1935 with *Collectivist Economic Planning: Critical Studies on the Possibilities of Socialism.* His fellow contributors included Ludwig von Mises (his mentor in Vienna), Professor N G Pierson (Amsterdam), Professor Georg Halm (Würzburg) and Professor Enrico Barone (Padua). His own contribution took up Mises's forceful discovery that the problem of knowing how best to use resources, faced by every socialist planner, was insuperable.

However, *Collectivist Economic Planning* did little to stem the tide of enthusiasm for Keynesian economics during the latter half of the 1930s. As one historian of the period puts it: 'That the Keynesian economic system was "politically" possible, and offered a painless solution to seemingly intractable problems, ensured its popularity; and all collectivists, Socialists, Liberals, and even many Conservatives, such as Macmillan, rushed to embrace it … to challenge Keynesianism was to be regarded as a reactionary and, in a favourite phrase of the time, a "die-hard".'[14]

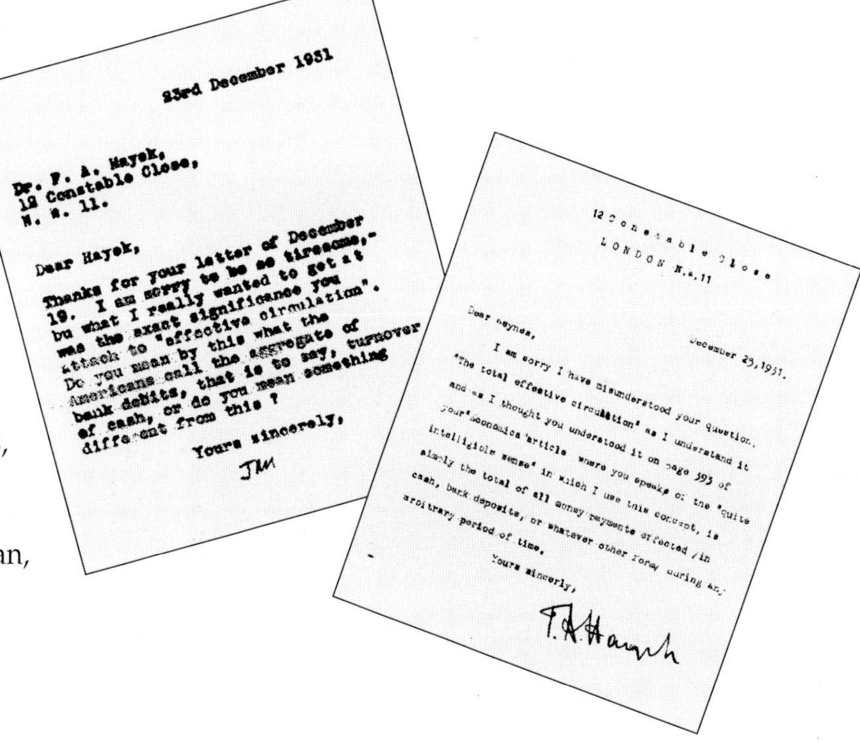

23rd December 1931

Dr. F. A. Hayek,
12 Constable Close,
N. W. 11.

Dear Hayek,

Thanks for your letter of December 19. I am sorry to be so tiresome,- but what I really wanted to get at was the exact significance you attach to "effective circulation". Do you mean by this what the Americans call the aggregate of bank debits, that is to say, turnover of cash, or do you mean something different from this?

Yours sincerely,
JMK

12 Constable Close
LONDON N.W.11

Dear Keynes,
December 23, 1931.

I am sorry I have misunderstood your question. "The total effective circulation" as I understand it and as I thought you understood it on page 393 of your Economica article where you speak of the "quite intelligible sense" in which I use this concept, is simply the total of all money payments effected (in cash, bank deposits, or whatever other form) during any arbitrary period of time.

Yours sincerely,
F.A.Hayek

Some of the correspondence (almost daily) between the two scholars in the early 1930s… discussing the definition of the 'effective circulation' of money.

Domestic life in England

In 1938, just a few months after the *Anschluss* when the Nazis invaded Austria and Hitler triumphantly entered Vienna, Hayek became a naturalized British subject.

The Times, August 13th, 1938
p. 8, under "Entertainments"!

Notice was given in last night's *London Gazette* that Mr. Paul Czinner, film producer, and husband of Elisabeth Bergner, and Dr. F. A. von Hayek, the Viennese professor, have been granted certificates of naturalization. Professor Hayek is Professor of Economic Science and Statistics in the University of London.

Hayek's note on this newspaper cutting shows his amusement that his naturalization announcement appeared in *The Times* under 'Entertainments'!

Hitler enters Vienna, 12 March 1938.

Hayek had become very contented with life in England. As he told *The Times* in 1985: *'Except for the Tyrolean mountains, all my emotional attachment is to England. I fell in love with England when I first went to Cambridge in January, 1931. Emotionally and intellectually it was my climate and it still is. It isn't really that the English are more intelligent than others, but they have great social strength.'*

During the 1930s, while busy writing and lecturing at the LSE, Hayek settled into his new life in London's leafy Hampstead Garden Suburb with his wife Hella and his two young children, Christine and Laurence.

'We lived very quietly with very little social life beyond the occasional entertaining of a visiting colleague. We were of course still running the house with the help of a regular maid. These were usually Austrian girls, one of whom stayed with us for a long time and became quite a member of the family. But this was about all that the then salary of a professor (at first £1,000 per annum, after five years £1,250) would support. Until 1936 we did without a car, and the one indulgence I granted myself was the membership in the Reform Club, which became very important to me.'[15]

'There were quite a group of the LSE economists living in the Garden Suburb at the time – apart from the Robbinses, who became our closest friends, Arnold Plant, Frank Paish, George Schwartz, and later for a time James Meade.'[16]

Christine and her mother outside their house in Constable Close, Hampstead Garden Suburb, in 1932.

Laurence, born 1934, with his mother.

Arthur Seldon, then a student at the LSE, recalls that, because Hayek's spoken English was so heavily accented, Plant took him for walks over Hampstead Heath to teach him English pronunciation.[17]

These days in London were interspersed with the equal contentment of Hayek's scholarly retreats to his beloved Austrian Alps. Here, he was returning to his roots. In these mountain forests, his father had studied and written about the local flora. Hayek would return to the mountains of Austria for inspiration throughout his life, saying that '*the combination of intellectual work and mountaineering fitted us well*'.

Hayek with his first car in the 1930s.

Hayek at St Jakob in 1933

The family relaxing happily in their garden in Hampstead.

Hayek re-examines the roots of capitalism

In his next book, *Profits, Interest and Investment*, published in 1939, Hayek defended the Austrian School's theory of the trade cycle and argued that monetary interventions cause far-ranging economic distortions that bring about malinvestments and unemployment. In the Preface, he said:

'The essays collected in this volume are a selection from the various attempts made in the course of the past ten years to improve and develop the outline of a theory of industrial fluctuations.... [The essays] contain points which I still feel are of some importance and since I do not yet feel ready to give a systematic exposition of the whole of this complex subject, to place these various attempts side by side within the covers of one volume is the best I can do to do justice to the many aspects of it.'

That *'systematic exposition'* would soon follow. Throughout the middle and later 1930s Hayek had been working on his reply to some of the objections raised against his earlier *Prices and Production*, especially its inadequate presentation of the theory of capital that it pre-supposed. In order to satisfy his critics, himself chief amongst them, Hayek decided to re-work his theory of *'capitalistic production'* from its foundation and with an intense and sustained effort nothing short of heroic.[19] This led to the publication in 1941 of *The Pure Theory of Capital* in both Britain and America.

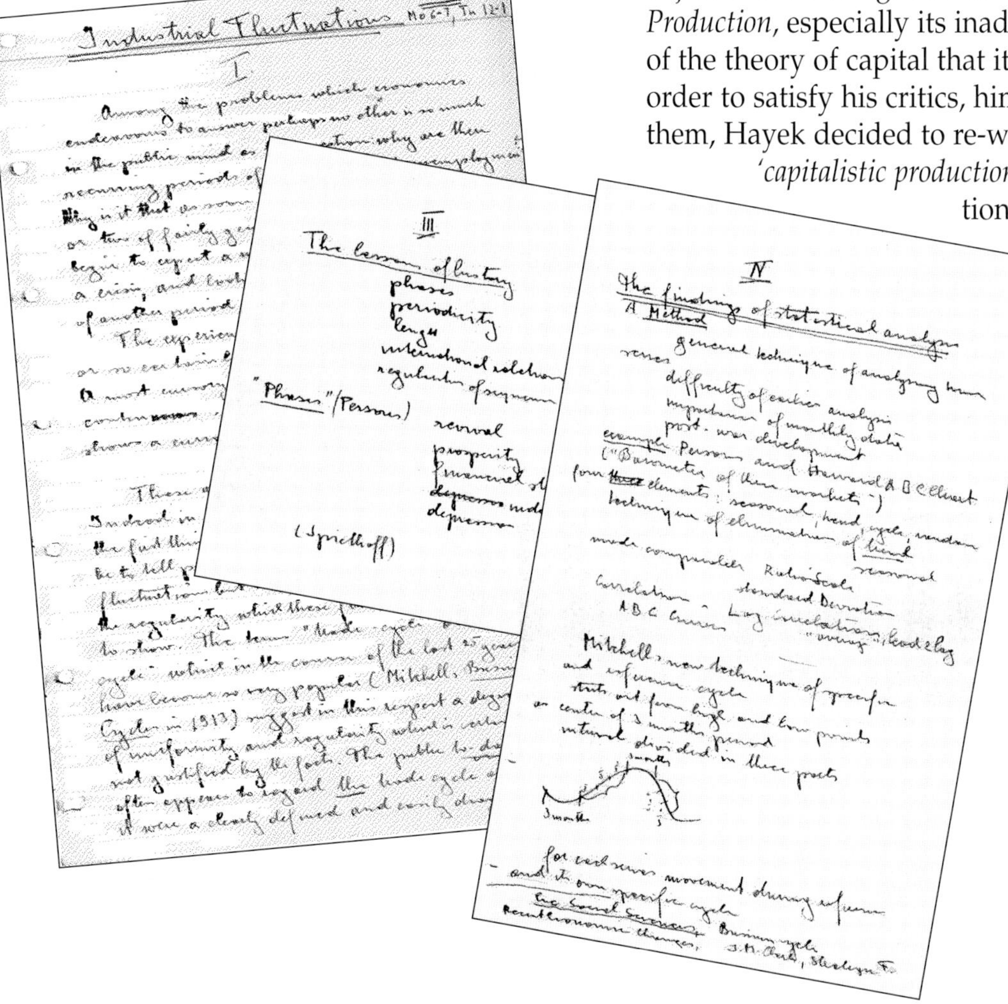
Industrial Fluctuations

I

III

The lessons of history

IV

The findings of statistical analysis

Some hand-written pages from Hayek's notebook for his 1938 LSE lectures on 'Industrial Fluctuations', which formed the basis for *Profits, Interest and Investment.*

THE
PURE THEORY
OF
CAPITAL

BY
FRIEDRICH A. HAYEK
TOOKE PROFESSOR OF ECONOMIC SCIENCE AND STATISTICS IN THE UNIVERSITY OF LONDON

LONDON
ROUTLEDGE & KEGAN PAUL LTD
BROADWAY HOUSE. 68–74 CARTER LANE, E.C.4

In the preface to this volume, Hayek wrote: '*This highly abstract study of a problem of pure economic theory has grown out of the concern with one of the most practical and pressing questions which economists have to face, the problem of the causes of industrial fluctuations. The attempt to elaborate a chain of reasoning which seems to throw important light on this question had made it painfully clear to me that some of the theoretical tools with which we are at present equipped are quite inadequate for the task.... [It therefore] seemed imperative... to turn back to the revision of the fundamentals and to work out a theory of capitalist production which would prove adequate for the analysis of dynamic changes.*'

However, Hayek then candidly admitted: '*My reluctance to undertake this work would have been greater if from the beginning I had been aware of the magnitude of the task that awaited me.*'

Henry Hazlitt (1894–1993), the leading American financial journalist, editorial columnist and long-time critic of Keynes.

In the United States, Henry Hazlitt reviewed the book in *Newsweek*, saying 'We cannot understand the causes and remedies of business cycles unless we understand the role played by capital. Hayek's book carries the analysis of the role and function of capital beyond the point to which it was brought by the great pioneers – notably Böhm-Bawerk, Jevons and Wicksell. It is the most penetrating study so far devoted to the subject.'

The point about *The Pure Theory of Capital* is that it looks under the surface of the averages and aggregates that were becoming the fashion among economists at the time, led of course by Keynes. By showing the very complex nature of capital and its importance in economic booms and slumps, the book stands as a classic in the field.

Hayek's rigorous approach was still being eclipsed by the popularity of the softer options proposed by Keynes. However, after the outbreak of war sent the staff and students of the London School of Economics to the University of Cambridge in 1939, Hayek would be given the opportunity to write one of the most influential (and controversial) books of the age.

6 *Hayek writes* The Road to Serfdom *in wartime Cambridge*

The quiet professor hits the headlines on both sides of the Atlantic, 1944–1945

After two decades of political and economic upheavals, Europe was plunged into a second world war in 1939.

The London School of Economics was evacuated to Peterhouse College, Cambridge, where Hayek continued his teaching and writing activities. Notwithstanding their scholarly disagreements, Keynes helped his friend Hayek obtain rooms at King's College, where Keynes was Bursar.

Britain's intellectuals continued to embrace planning, even as the Nazi version of state socialism was bombing their country. This prompted Hayek to write a short book, *The Road to Serfdom*, as a stark warning explaining how even democratic socialism could evolve all too easily into totalitarianism. Unexpectedly successful, the book then threw him into the public and political limelight, though it discredited him among most of his academic contemporaries.

In America, his book made the non-fiction best sellers' list almost at once, bringing him praise and condemnation in almost equal measure.

A masterly condensation of *The Road to Serfdom*, published in 1945 in the *Reader's Digest*, brought Hayek's ideas to the attention of thousands of people, especially in the United States, where his 1945 lecture tour attracted audiences in unprecedented numbers.

This celebrity would pave the way for another dramatic change in his career a few years later, when he was invited to the University of Chicago in 1950. But by then he had already proposed a simple, but in the event a very successful, solution that would help preserve freedom in the aftermath of the destructive world war that had just come to an end.

Sandbagged buildings in King's Parade, Cambridge.

Hayek's wartime interlude in Cambridge

After the Second World War broke out in September 1939, the London School of Economics was evacuated to the University of Cambridge. There, at the invitation of the Master and Fellows, the LSE was accommodated at Peterhouse College until 1945. For a time, Hayek was the only senior LSE economist, because many of his British-born colleagues, such as Robbins, entered the government's war service as civil servants.

Hayek initially went to Cambridge from London for only three days a week, but with the beginning of the bombing in September 1940, this became impracticable. At first, Hayek could not get suitable accommodation in Cambridge; so the Robbinses, who at that time had a cottage in the Chilterns, took his family for a year, while Keynes found him rooms at King's College.

2 INDEPENDENT PRESS AND CHRONICLE, Friday, June 21, 1940.

TUESDAY'S RAID IN CAMBRIDGESHIRE

Photo] The wreckage of part of the eight Cambridgeshire houses. [Press & Chronicle

Casualties in One Locality

There was no sign whatsoever of any panic. A.R.P. services went into operation smoothly and efficiently.

The A.F.S. speedily dealt with small outbreaks of fire and rescue and demolition parties removed most of the injured in a very short

in a semi-conscious condition. A tunnel was made beneath the rubbish and a soldier crawled into it and held the chair while workers made frantic efforts to fasten a rope around it. He was there for three hours before

Bombs fell in Cambridgeshire too, as the local newspaper's cuttings show.

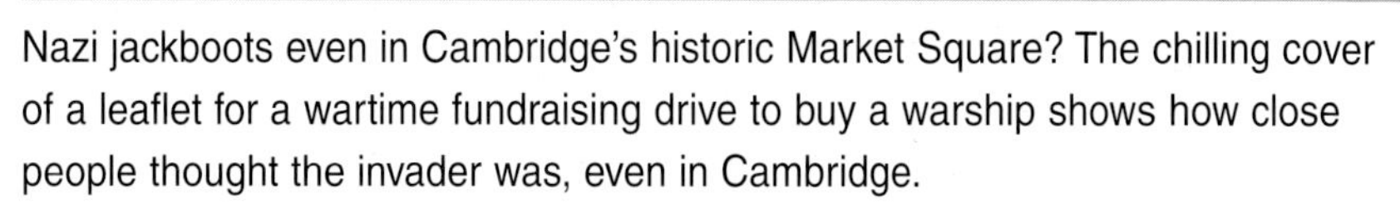

Nazi jackboots even in Cambridge's historic Market Square? The chilling cover of a leaflet for a wartime fundraising drive to buy a warship shows how close people thought the invader was, even in Cambridge.

Vehicles in the same Market Square, Cambridge.

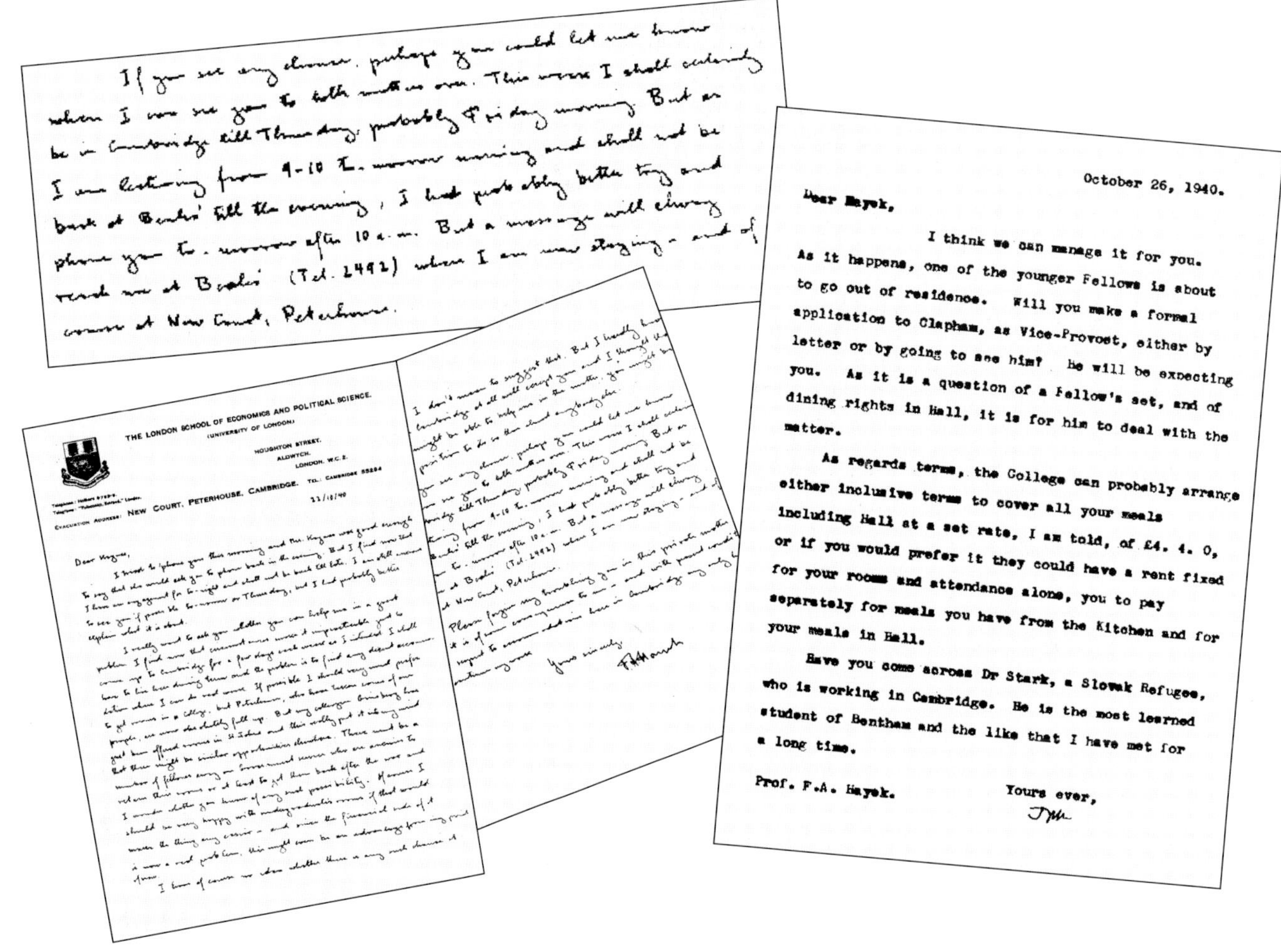

October 26, 1940.

Dear Hayek,

I think we can manage it for you. As it happens, one of the younger Fellows is about to go out of residence. Will you make a formal application to Clapham, as Vice-Provost, either by letter or by going to see him? He will be expecting you. As it is a question of a Fellow's set, and of dining rights in Hall, it is for him to deal with the matter.

As regards terms, the College can probably arrange either inclusive terms to cover all your meals including Hall at a set rate, I am told, of £4. 4. 0, or if you would prefer it they could have a rent fixed for your rooms and attendance alone, you to pay separately for meals you have from the Kitchen and for your meals in Hall.

Have you come across Dr Stark, a Slovak Refugee, who is working in Cambridge. He is the most learned student of Bentham and the like that I have met for a long time.

Prof. F.A. Hayek.

Yours ever,

JMK

Some of the correspondence from 1940, when Keynes helped Hayek settle in Cambridge.

Hayek felt that the LSE students greatly benefited from the joint Cambridge–London economics teaching. This task he shared with Professor A C Pigou of King's College, who had been appointed in 1908 to Marshall's Chair of Political Economy as his chosen successor.

Pigou had defended Hayek against Keynes's 'bodyline bowling' in their dispute over *A Treatise on Money.* Hayek gave the lectures on advanced economic theory for both institutions, while Pigou gave the elementary theory. And as well as economics, Hayek and Pigou shared the same enthusiasm for mountaineering.

The great Cambridge economist A C Pigou (1877–1959).

Hayek is seen here with some of his LSE students at Grove Lodge, Cambridge, 1942. Seated second right is Arthur Lewis, who became the 1979 Nobel Laureate in Economics.

Hayek looked back on his years at Cambridge with great affection. '*Life at Cambridge during those war years was to me particularly congenial, and it completed the process of thorough absorption in English life which, from the beginning, I had found very easy.... And of all the forms of life, that at one of the colleges of the old universities – at least as it then was at Cambridge – still seems to me the most attractive. The evenings at the High Table and the Combination Room at King's are among the pleasantest recollections of my life.*'[1]

Yet Hayek's scholarly life in Cambridge contrasted with the worsening reports of the war in continental Europe, as the Nazi war machine engulfed more and more countries. By 1942, the Axis powers occupied most of Europe.

The quiet professor, awarded a DSc by the University of London in 1943, would soon make his own, intellectual, entry into the fray – and in consequence would find himself in the headlines in both Europe and the United States.

Malting Lane, Cambridge: Hayek lived here at the Old Oast House with his family from 1941 until 1945. It was here that he drafted *The Road to Serfdom*.

Hayek writes *The Road to Serfdom*

During the war years, Britain's socialist intellectuals found the famous weekly *Picture Post* a highly effective vehicle for reaching a mass audience with articles on the need for planning in many aspects of life.

PICTURE POST

MR. KEYNES HAS A PLAN

THE NEW BRITAIN MUST BE PLANNED

by Maxwell Fry

PICTURE POST

BEVERIDGE: THE FIGHT IS ON

Victor Gollancz's Left Book Club also provided a popular platform for political activists. Meanwhile, Sir William Beveridge, Master of University College Oxford since 1937, was advancing the case for greater state intervention with his sensational *Report on Social Insurance and Allied Services* (1942) and its sequel *Full Employment in a Free Society* (1944), which between them laid the foundations of the postwar welfare state in Britain.

Increasing public support for planning, the wide political attraction of communism for many in the 1930s and the destruction brought about by Nazi aggression all deeply depressed Hayek. To warn people of how the 'planning' ideal could go wrong and so easily turn into a totalitarian nightmare, he decided to write a short book for non-academic readers, which he dedicated *'To the Socialists of All Parties'*.

'The light burden of teaching (there were very few students) and the short distances at Cambridge gave me more time for my own work than I ever had before. Though my main interest was still in pure economic theory, it was at Cambridge that I wrote The Road to Serfdom *....'*[2]

Turned down for any war effort because of his Austrian origins (even though by then he was a naturalized British subject), Hayek saw his work on the book *'as a duty which I must not evade'*.

Hayek's typewriter.

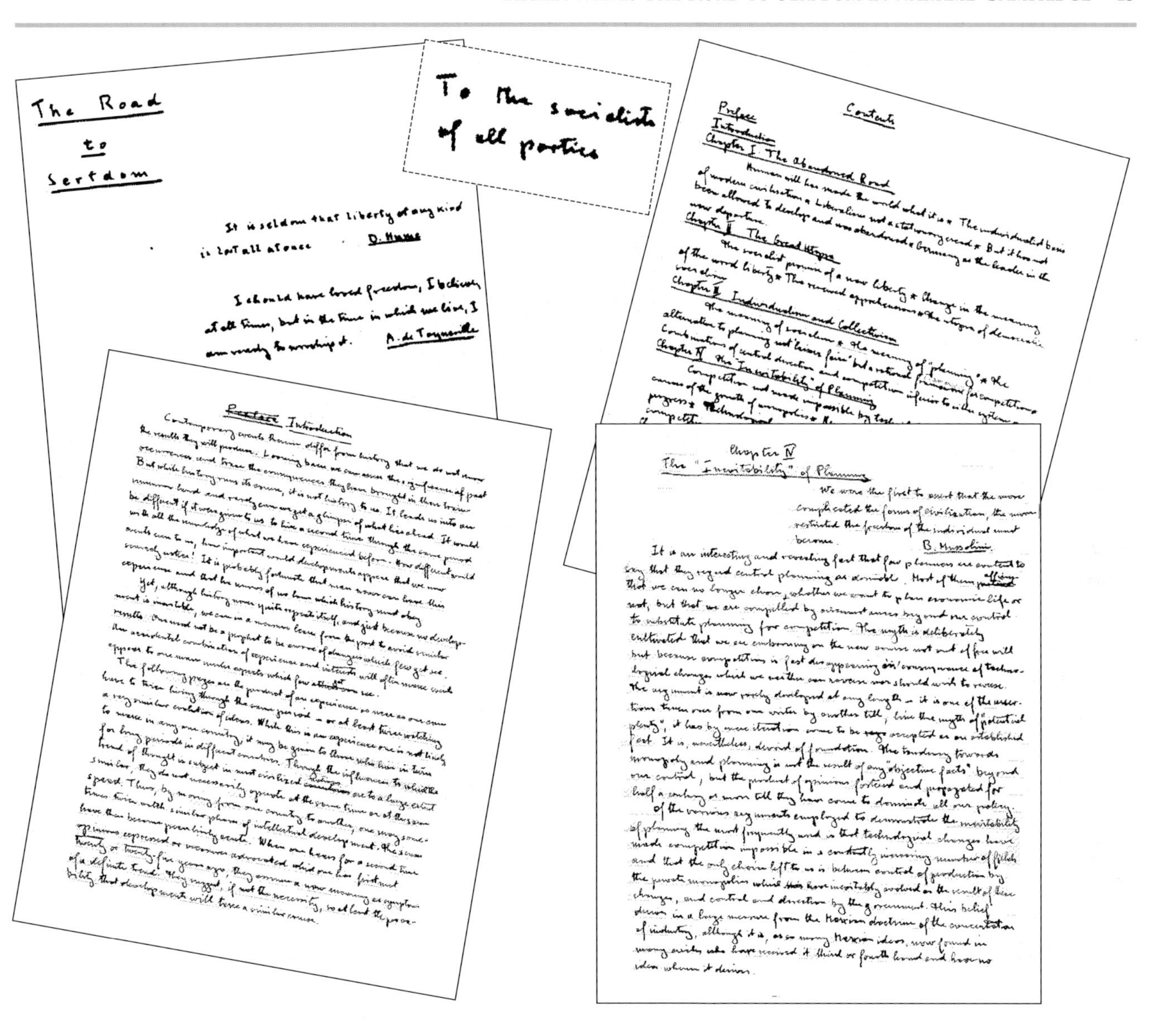

The Road to Serfdom

To the socialists of all parties

It is seldom that liberty of any kind is lost all at once. D. Hume

I should have loved freedom, I believe, at all times, but in the time in which we live, I am ready to worship it. A. de Tocqueville

Contents

Preface

Introduction

Chapter I The Abandoned Road

Chapter II The Great Utopia

Chapter III Individualism and Collectivism

Chapter IV The Inevitability of Planning

Introduction

Chapter IV

The "Inevitability" of Planning

We were the first to assert that the more complicated the forms of civilization, the more restricted the freedom of the individual must become. B. Mussolini

The Road to Serfdom in its third draft – hand written by Hayek in a school notebook.

The Road to Serfdom, published on 10 March 1944, is a devastating demonstration of how even democratic socialism can become subverted into totalitarianism. Not surprisingly, it immediately attracted headlines in the British daily and weekly press with reviews and comments ranging from what Hayek called *'lavish praise'* to *'passionate hatred'*.[3]

The book was quoted in Parliament; *Time and Tide* thought it 'enthralling and disturbing'; *The Scotsman* said it dealt with 'questions of the highest importance to every citizen interested in the future of his country'. Predictably, however, the *Manchester Guardian* accused him of 'looking backwards'; while the *New Statesman* said his views only had 'a curiosity interest'. By contrast, *The Spectator* said that Hayek's book had touched upon 'most fateful questions'.

that will shape policy

PLEA FOR LIBERALISM

THE ROAD TO SERFDOM.—By F. A. Hayek. (10s. 6d. Routledge.)

d to Serfdom

HAYEK

OF THE DAY

MANCHESTER GUARDIAN
April 14th, 1944

e Socialist Error

LOOKING BACKWARD

ATTACK ON "PLANNING"

An Outstanding Book

B.D.J.

PROFESSOR HAYEK'S book "The Road to Serfdom" is the most important of its kind

GROUNDS FOR DISMAY

The Road to Serfdom. By F. A. Hayek. (Routledge. 10s. 6d.)
The Mirror of the Past. By K. Zilliacus. (Gollancz. 7s. 6d.)

By GEORGE ORWELL

Back to Adam Smith

THE LISTENER
30 March 1944

The Road to Serfdom. By F. A. Hayek. Routledge. 10s. 6d.

THIS BOOK SHOULD BE READ by everybody.

THE TIMES
15 APR 1944

THE NEW STATESMAN AND NATION
May 13th, 1944

THE CASE AGAINST SOCIALISM

ECONOMIC JOURNAL

REVIEWS

FREEDOM AND PLANNING

CASE FOR THE INDIVIDUALIST

THE ROAD TO SERFDOM. By F. A. HAYEK. Routledge.

As its title suggests, this powerful statement of the individualist case argues that collectivist planning—distinct from liberal planning to remove restrictions—must lead to totalitarianism and so to servitude.

NEW WORLDS

A. C. PIGOU

SUNDAY TIMES
9 April 1944

TIME AND TIDE
June 10th, 1944

berties

YOUNG

LIBERAL MAGAZINE
April 1944

FRUITGROWER
April 13, 1944

—THE SPECTATOR, JULY 7, 1944

MARGINAL COMMENT

By HAROLD NICOLSON

NEWS & BOOK TRADE
March 18, 1944

TIME AND TIDE

From Planning to Gallows

"THE ROAD TO SERFDOM." By Prof. F. A. Hayek. (London: George Routledge & Sons, Ltd.; 10s. 6d. net.)

Books people are reading

The Financial News devoted its first leader to his 'Attack on Planning' and the influential BBC publication *The Listener* said it 'should be read by everybody'. The BBC broadcast a lengthy summary; Michael Polyani in *The Times Literary Supplement* said it was a 'powerful statement' of the individualist case against collectivist planning, while even *The Fruitgrower*, the *Hardware Trade Journal* and *The Electrician* all carried reviews!

Hayek's friend Pigou thought it a 'scholarly and sincere book' while George Orwell found its effect 'depressing'. The blunt-speaking *Yorkshire Post* said Hayek had delivered 'a frontal attack on that conception of a "planned economy" which nowadays commands general assent from so many quarters'. As a result of all the publicity, Routledge's initial print run of 2,000 was sold out within a few days. Reprint followed reprint, despite wartime paper rationing in Britain.

A publicity photograph of Hayek taken in 1944 by the famous Cambridge photographer, Lettice Ramsay, for the launch of *The Road to Serfdom* by Routledge.

Keynes reviews *The Road to Serfdom*

J M Keynes (right) at Bretton Woods with Henry Morgenthau, President Roosevelt's Treasury Secretary.

Hayek's friendship with Keynes had flourished at Cambridge, and Keynes reviewed *The Road to Serfdom* in his famous long letter of 28 June 1944 to Hayek, written *en route* to the Bretton Woods monetary conference in America.

Although Keynes wrote expansively: 'In my opinion it is a grand book.... I find myself in agreement with virtually the whole of it', he was clearly *not* convinced by some of Hayek's arguments, adding later: 'I should therefore conclude your theme rather differently. I should say that what we want is not no planning, or even less planning, indeed I should say that we almost certainly want more.'[4]

Claridge Hotel,
Atlantic City, N.J.
28th June 1944.

My dear Hayek,

The voyage has given me the chance to read your book properly. In my opinion it is a grand book. We all have the greatest reason to be grateful to you for saying so well what needs so much to be said. You will not expect me to accept quite all the economic dicta in it. But morally and philosophically I find myself in agreement with virtually the whole of it; and not only in agreement with it, but in a deeply moved agreement.

Turning to a few special points, I think you strike ... wrong note on Page 69 where you deprecate all the talk ... round the corner. No doubt this is partly ... view to yours about the facts. ... in line with your ...

Prof. F.A.Hayek,
The Old Oast House,
Malting Lane,
Cambridge.

-2-

They ask us to concentrate on economic conditions more exclusively than in any earlier period in the world's history precisely at the moment when by their own showing technical achievement is making this sacrifice increasingly unnecessary. This preoccupation with the economic problem is brought to its most intense at a phase in our evolution when it is becoming ever less necessary.

The line of argument you yourself take depends on the very doubtful assumption that planning is not more efficient. Quite likely from the purely economic point of view it is efficient. That is why I say that it would be more in line with your general argument to point out that ...

-3-

I come finally to what is really my only serious criticism of the book. You admit here and there that it is a question of knowing where to draw the line. You agree that the line has to be drawn somewhere, and that the logical extreme is not possible. But you give us no guidance whatever as to where to draw it. In a sense this is shirking the practical issue. It is true that you and I would probably draw it in different places. I should guess that according to my ideas you greatly underestimate the practicability of the middle course. But as soon as you admit that the extreme is not possible, and that a line has to be drawn, you are, on your own argument, done for, since you are trying to persuade us that so soon as one moves an inch in the planned direction you are necessarily launched on the slippery path which will lead you in due course over the precipice.

I should therefore conclude your theme rather differently. I should say that what we want is not no planning, or even less planning, indeed I should say that we almost certainly want more. But the planning should take place in a community in which as many people as possible, both leaders and followers, wholly share your own moral position. Moderate planning will be safe if those carrying it out are rightly orientated in their own minds and hearts to the moral issue. This is in fact already true of some

-4-

of them. But the curse is that there is also an important section who could almost be said to want planning not in order to enjoy its fruits but because morally they hold ideas exactly the opposite of yours, and wish to serve not God but the devil. Reading the New Statesman & Nation one sometimes feels that those who write there, while they cannot safely oppose moderate planning, are really hoping in their hearts that it will not succeed; and so prejudice more violent action. They fear that if moderate measures are sufficiently successful, ...

-5-

issues. Dangerous acts can be done safely in a community which thinks and feels rightly, which would be the way to hell if they were executed by those who think and feel wrongly.

Yours ever,

K.

I should therefore conclude your theme rather differently. I should say that what we want is not no planning, or even less planning, indeed I should say that we almost certainly want more.

The British Academy admits Hayek

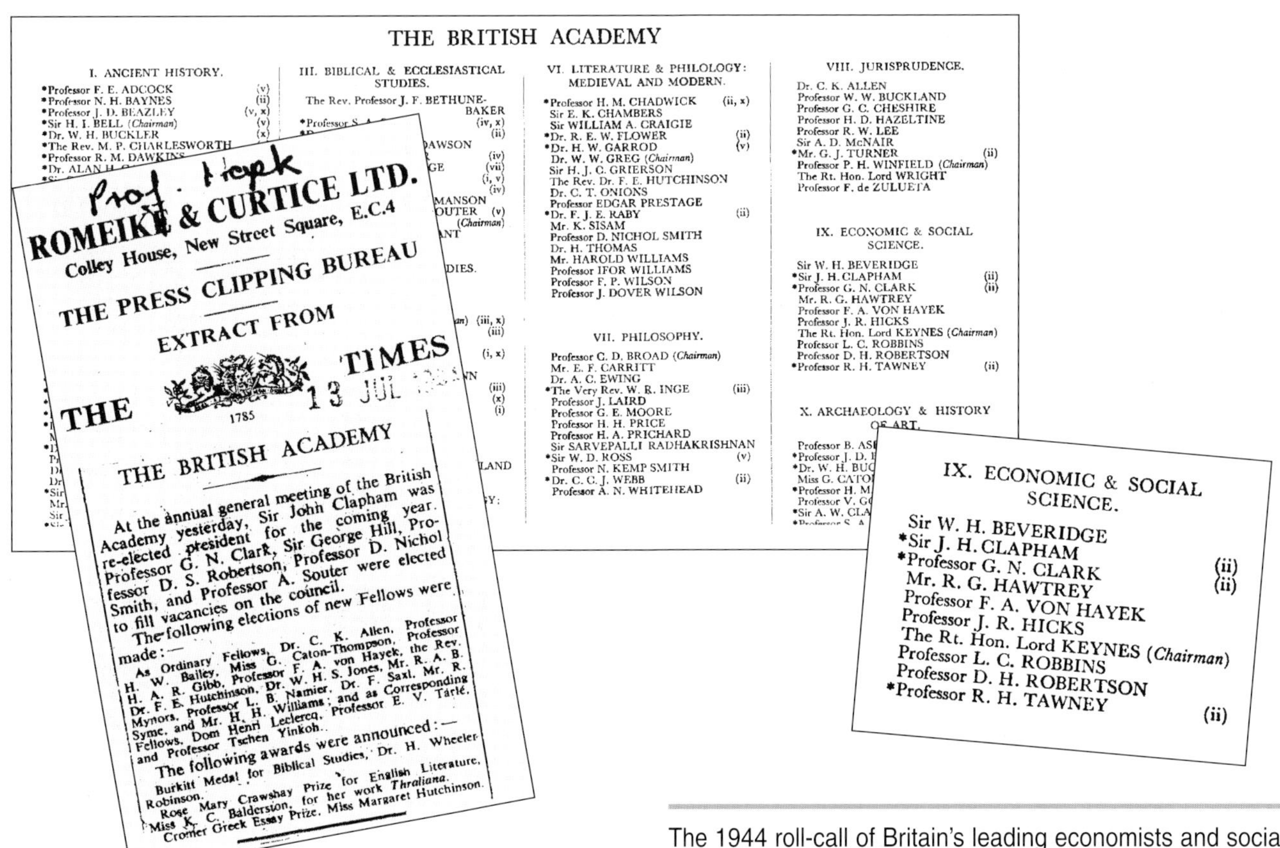

THE BRITISH ACADEMY

I. ANCIENT HISTORY.

*Professor F. E. ADCOCK (v)
*Professor N. H. BAYNES (ii)
*Professor J. D. BEAZLEY (v, x)
*Sir H. I. BELL (*Chairman*) (v)
*Dr. W. H. BUCKLER (x)
*The Rev. M. P. CHARLESWORTH
*Professor R. M. DAWKINS

III. BIBLICAL & ECCLESIASTICAL STUDIES.

The Rev. Professor J. F. BETHUNE-BAKER

VI. LITERATURE & PHILOLOGY: MEDIEVAL AND MODERN.

*Professor H. M. CHADWICK (ii, x)
Sir E. K. CHAMBERS
Sir WILLIAM A. CRAIGIE
*Dr. R. E. W. FLOWER (ii)
*Dr. H. W. GARROD (v)
Dr. W. W. GREG (*Chairman*)
Sir H. J. C. GRIERSON
The Rev. Dr. F. E. HUTCHINSON
Dr. C. T. ONIONS
Professor EDGAR PRESTAGE
*Dr. F. J. E. RABY (ii)
Mr. K. SISAM
Professor D. NICHOL SMITH
Dr. H. THOMAS
Mr. HAROLD WILLIAMS
Professor IFOR WILLIAMS
Professor F. P. WILSON
Professor J. DOVER WILSON

VII. PHILOSOPHY.

Professor C. D. BROAD (*Chairman*)
Mr. E. F. CARRITT
Dr. A. C. EWING
*The Very Rev. W. R. INGE (iii)
Professor J. LAIRD
Professor G. E. MOORE
Professor H. H. PRICE
Professor H. A. PRICHARD
Sir SARVEPALLI RADHAKRISHNAN
*Sir W. D. ROSS (v)
Professor N. KEMP SMITH
*Dr. C. C. J. WEBB (ii)
Professor A. N. WHITEHEAD

VIII. JURISPRUDENCE.

Dr. C. K. ALLEN
Professor W. W. BUCKLAND
Professor G. C. CHESHIRE
Professor H. D. HAZELTINE
Professor R. W. LEE
Sir A. D. McNAIR
*Mr. G. J. TURNER (ii)
Professor P. H. WINFIELD (*Chairman*)
The Rt. Hon. Lord WRIGHT
Professor F. de ZULUETA

IX. ECONOMIC & SOCIAL SCIENCE.

Sir W. H. BEVERIDGE
*Sir J. H. CLAPHAM (ii)
*Professor G. N. CLARK (ii)
Mr. R. G. HAWTREY
Professor F. A. VON HAYEK
Professor J. R. HICKS
The Rt. Hon. Lord KEYNES (*Chairman*)
Professor L. C. ROBBINS
Professor D. H. ROBERTSON
*Professor R. H. TAWNEY (ii)

X. ARCHAEOLOGY & HISTORY OF ART.

Prof. Hayek

ROMEIKE & CURTICE LTD.
Colley House, New Street Square, E.C.4
THE PRESS CLIPPING BUREAU
EXTRACT FROM
THE TIMES
1785
13 JUL

THE BRITISH ACADEMY

At the annual general meeting of the British Academy yesterday, Sir John Clapham was re-elected president for the coming year. Professor G. N. Clark, Sir George Hill, Professor D. S. Robertson, Professor D. Nichol Smith, and Professor A. Souter were elected to fill vacancies on the council.

The following elections of new Fellows were made:—

As Ordinary Fellows, Dr. C. K. Allen, Professor H. W. Bailey, Miss G. Caton-Thompson, Professor H. A. R. Gibb, Professor F. A. von Hayek, the Rev. Dr. F. E. Hutchinson, Dr. W. H. S. Jones, Mr. R. A. B. Mynors, Professor L. B. Namier, Dr. F. Saxl, Mr. R. Syme, and Mr. H. H. Williams; and as Corresponding Fellows, Dom Henri Leclercq, Professor E. V. Tarlé, and Professor Tschen Yinkoh.

The following awards were announced:—Burkitt Medal for Biblical Studies, Dr. H. Wheeler Robinson. Rose Mary Crawshay Prize for English Literature, Miss K. C. Balderston, for her work *Thraliana*. Cromer Greek Essay Prize, Miss Margaret Hutchinson.

The 1944 roll-call of Britain's leading economists and social scientists.

At about the same time, Hayek was elected as a Fellow of the British Academy, proposed by Keynes. The timing was lucky: Sir John Clapham, (Vice-Provost of King's College, Cambridge), told Hayek that if the publication of *The Road to Serfdom* had been any earlier, he would never have been elected a member, such was the intellectual tenor of the times.

Now officially a member of the elite of British social scientists, Hayek had come a long way since that first London conference in the 1920s when he was the young director of a newly formed and obscure Austrian economic research institute in Vienna.

Hayek's sensational effect in the United States

After *The Road to Serfdom* had been published in the UK by Routledge, it was then published (at the suggestion of Fritz Machlup, Frank Knight and Aaron Director) by the University of Chicago Press in September 1944. Several other publishers had already turned it down: one even returned it as '*unfit for publication by a reputable house*', according to Hayek in the Preface to the 1956 edition.

Overnight, *The Road to Serfdom* became a best-seller in the United States. The University of Chicago Press had to reprint it three times in less than three weeks, during which time 17,000 copies had been produced.[5]

The prestigious *New York Times Book Review* of 24 September 1944 carried a front page review by Henry Hazlitt. Other reviews flowed in. Graham Hutton featured it in the *Chicago Daily News* and the reviewer in a San Antonio paper agreed it was 'one of the great books of the century'. Coast-to-coast press coverage produced headlines ranging from 'Central Planning Leads to Slavery' (*St Louis Post-Dispatch*) to 'Must we go Totalitarian?' (*Hartford Courant*). The lead editorial in *The New York Mirror* on 25 May 1945 said that '*The Road to Serfdom* is a book that may be placed among the greatest ever written on what confronts Americans today – LIBERTY'.

By contrast, the American arch-Keynesian, Professor Alvin Hansen of Harvard, wrote in the *New Republic* that: 'This kind of writing is not scholarship. It is seeing hobgoblins under every bed.'

The New York Times Book Review

SEPTEMBER 24, 1944 — Copyright, 1944, by The New York Times Company — VOL. XLIX—No. 39

AN ECONOMIST'S VIEW OF 'PLANNING'

Regimentation on the Fascist Model, Says Dr. Hayek, Can Evolve From It

THE ROAD TO SERFDOM. By Friedrich A. Hayek. With a Foreword by John Chamberlain. 250 pp. Chicago: University of Chicago Press. $2.75.

By HENRY HAZLITT

IN "The Road to Serfdom" Friedrich A. Hayek has written one of the most important books of our generation. It restates for our time the issue between liberty and authority with the power and rigor of reasoning that John Stuart Mill stated the issue for his own generation in his great essay, "On Liberty." It throws a brilliant light along the direction in which the world has been heading, first slowly, but now at an accelerative rate, for the last half-century. It is an arresting call to all well-intentioned planners and socialists, to all those who are sincere democrats and liberals at heart, to stop, look and listen.

Mr. Hayek, an internationally known economist, is a native of Austria. He was director of the Austrian Institute for Economic Research and lecturer in economics at the University of Vienna during the years in which Hitler was building up his instruments to seize power. He has lived in England since 1931, however, when he became Professor of Economic Science at the University of London. In 1938 he became a British citizen.

This experience has given him a type of insight not granted to the usual American or English economist or student of political affairs. "The following pages," as he writes, "are the product of an experience as near as possible to twice living through the same period. * * * When one hears for a second time opinions expressed or measures advocated which one has first met twenty or twenty-five years ago, they assume a new meaning as symptoms of a definite trend. They suggest, if not the necessity, at least the probability, that developments will take a similar course. It is necessary now to state the unpalatable truth that it is Germany whose fate we are in some danger of repeating."

an example to be imitated, just as in more recent years Sweden has been the model country to which progressive eyes were directed.

The supreme tragedy is still not seen that in Germany it was largely people of good-will, men who were admired and held up as models in the democratic countries, who prepared the way for, if they did not actually create, the forces which now stand for everything they detest. Few are ready to recognize that the rise of fascism and nazism was not a reaction against the Socialist trends of the preceding period but a necessary outcome of those tendencies. This is a truth which most people were unwilling to see even when the similarities of many of the repellent features of the internal regimes in communist Russia and National Socialist Germany were widely recognized. As a result many who think themselves infinitely superior to the aberrations of nazism, and sincerely hate all its manifestations, work at the same time for ideals whose realization would lead straight to the abhorred tyranny.

MR. HAYEK's book is so compactly written that it is impossible adequately to summarize the closely woven argument by which he seeks to prove this thesis. He shows how the very success of liberalism in the nineteenth century became the cause of its decline. The miracles of production and social advance to which it led caused the world to take this enormous progress for granted, to forget the policy of freedom on which it rested, and to become impatient because the unprecedented pace of progress was not still greater. Although we had been warned by some of the greatest political thinkers of the nineteenth century, by de Tocqueville and Lord Acton, that socialism means slavery, we have so completely forgotten the warning that, even when this new form of slavery rises before our eyes, it scarcely occurs to us that the two things may be connected.

Mr. Hayek points out how England, the great center from which for more than two hundred years the world had

Hayek's status as an international figure was to be vastly enhanced by the publication of a condensation of his book in the *Reader's Digest*. It gave *The Road to Serfdom* much wider publicity than it could ever have received on its own. Hayek thought the condensation was *'very well done,'* although he *'did not get a penny'* from it.[6] Even so, it proved an effective way to spread the ideas of his book in the United States, particularly when the Book of the Month Club then distributed 600,000 copies of this condensed version.

Mises in his study in New York in the late 1940s.

His old colleague and friend, Ludwig von Mises, now teaching at New York University, wrote to Hayek on 23 February 1945 to say: 'The news of your impending lecture tour is very gratifying. It is almost a public sensation. You probably do not realize how great the success of your book is and how popular you are in this country.'[7]

'The popular success of The Road to Serfdom *was a complete surprise to me. Though I long resisted the pull which threatened to draw me from pure theory into more practical work, it had ultimately a profound effect on my life.... [F]rom March to May 1945, just when the book made the best-seller list, I was on [an academic] lecture tour in the USA, which was one of the most curious experiences of my life....*

'While I was crossing the Atlantic in slow convoy (it was still during the war) and without communications, the condensation of the book in the Reader's Digest *completely altered the position. I was suddenly if only temporarily famous, and on arrival was told that the whole plan for my visit was changed, that I was to go on a far-ranging popular lecture tour, and that all arrangements had been put in the hands of a commercial lecture agency.'*[8]

The University of Chicago Press printed 10,000 copies of a pocket-size version, a seventh impression, ready for Hayek's 1945 speaking tour in the United States. This generated even more massive press coverage for the book, which in turn initiated heated discussion in economic and political circles. And for every person who read his actual book, thousands more read the *Reader's Digest* condensation and the press features about it.

'The extraordinary success of my book was quite unexpected to me,' wrote Hayek. But he was still not optimistic about the future: *'The prospects for Europe seem to me as dark as possible.'*[9]

The READER'S DIGEST

An article a day of enduring significance, in condensed permanent booklet form

April 1945

THE ROAD
TO
SERFDOM

A CONDENSATION FROM THE BOOK BY

Friedrich A. Hayek

"IN *The Road to Serfdom*," writes Henry Hazlitt in the New York *Times*, "Friedrich A. Hayek has written one of the most important books of our generation. It restates for our time the issue between liberty and authority. It is an arresting call to all well-intentioned planners and socialists, to all those who are sincere democrats and liberals at heart, to stop, look and listen."

The author is an internationally known economist. An Austrian by birth, he was director of the Austrian Institute for Economic Research and lecturer in economics at the University of Vienna during the years of the rise of fascism in Central Europe. He has lived in England since 1931, when he became Professor of Economic Science at the University of London, and is now a British citizen.

Professor Hayek, with great power and rigor of reasoning, sounds a grim warning to Americans and Britons who look to the government to provide the way out of all our economic difficulties. He demonstrates that fascism and what the Germans correctly call National Socialism are the inevitable results of the increasing growth of state control and state power, of national "planning" and of socialism.

In a foreword to *The Road to Serfdom* John Chamberlain, book editor of *Harper's*, writes: "This book is a warning cry in a time of hesitation. It says to us: Stop, look and listen. Its logic is incontestable, and it should have the widest possible audience."

Copyright 1944, The University of Chicago, and published at $2.75 by the University of Chicago Press, Chicago 37, Ill.

SUNDAY GRAPHIC, June 17, 1945.

The Road To Serfdom

Exclusive Extracts From The Remarkable Book Everyone Is Talking About

By PROFESSOR F. HAYEK

I HAVE spent about half my adult life in close touch with German thought, and the other half in the United States and Britain.

In the latter period I have become increasingly convinced that some of the forces which destroyed freedom in Germany are also at work here.

Let us remember that 15 years ago the possibility of such a thing happening in Germany would have appeared just as fantastic not only to nine-tenths of the Germans themselves but also to the most hostile foreign observer.

The Supreme Tragedy

THERE are many features then regarded as "typically German" which are now equally familiar in America and England: the increasing veneration for the State, the fatalistic acceptance of "inevitable trends," the enthusiasm for "organisation" of everything (we now call it "planning")...

plan, it is not merely transformed but infinitely heightened.

In every real sense a badly-paid unskilled workman in this country has more freedom to shape his life than many an employer in Germany or a much better paid engineer or manager in Russia.

If he wants to change his job or the place where he lives, if he wants to profess certain views or spend his leisure in a particular way, he faces no absolute impediments.

Our generation has forgotten that the system of private property is the most important guarantee of freedom.

It has been well said that, in a country where the sole employer is the State, opposition means death by slow starvation.

Killed By Socialism

It is disquieting to see in England and the United States to-day the same drawing together of forces and nearly the same contempt of all that is liberal in the old sense.

"Planning" owes its popularity largely to the fact that everybody desires, of course, that we should handle our common problems with as much foresight as possible.

The question is whether we should create conditions under which the knowledge and initiative of individuals are given the best scope so that *they* can plan most successfully; or whether we should direct and organise all economic activities according to a "blueprint."

The successful use of competition does not preclude some types of Government interference; to limit working hours, to require certain sanitary arrangements, to provide an extensive system of social services is fully compatible with the preservation of competition.

Planning and competition can be combined only by planning *for* competition, not by planning *against* competition.

What is promised to us as the Road to Freedom is in fact the Highroad to Servitude.

For it is not difficult to see what must be the consequences when de-

What Attlee Said

"THE ROAD TO SERFDOM" became the nation's No. 1 talking point when Mr. Attlee referred to it in his election broadcast, saying that the theories of its author, Professor Hayek, "were popular just now among Conservatives."

This was, to put it mildly, an understatement. Professor Hayek, who occupies the Chair of Economics at London University, has become a best-selling author in two continents since "The Road to Serfdom" was first published here and in America.

It is temporarily unobtainable, but the "Sunday Graphic" begins to-day a series of exclusive extracts, the second of which will appear next week.

Winston (later Sir Winston) Churchill (1875–1965), Britain's wartime Prime Minister.

Over the next half century, *The Road to Serfdom* sold millions worldwide in its various editions, with the University of Chicago Press alone selling 250,000 copies. Authorised editions were to be published in Australia (1944), Sweden (1944), France (1945), Germany (1945), Denmark (1946), Portugal (1946), Spain (1946), the Netherlands (1948), Italy (1948), Switzerland (1948), Norway (1949), Japan (1954), Taiwan (1956), Iceland (1980) and Russia (1983).

Taking into account the millions of book sales in some 20 countries, the *Reader's Digest* and the Book of the Month Club condensations (to say nothing of 'underground' editions in countries behind the Iron Curtain), and the hundreds of column inches of press coverage in the UK, the United States and elsewhere, no economist before or since Friedrich Hayek has ever reached such a worldwide audience.

Yet, despite his wide press and public acclaim in the United States, Hayek's decline into academic purdah accelerated. With the publication of *The Road to Serfdom*, Hayek admitted, '*I had made myself very unpopular with intellectuals with my attack on Socialism*'.[10]

Hayek was to hit the headlines again in the British press, when he actually became an issue in the British general election of 1945. Churchill, in a BBC election campaign broadcast on 4 June, presumably with *The Road to Serfdom* in mind, warned his listeners that 'Socialism is inseparably woven with totalitarianism and the abject worship of the state'. He then said that if a Labour government committed itself to carrying out its socialist programme, it would have to fall back on some form of Gestapo to police it.

Churchill's remark caused a furore, and Hayek's views were drawn into the argument when the Labour leader, Clement Attlee, linked him to Churchill as an 'adviser'. In fact, though Churchill knew of *The Road to Serfdom*, Hayek had met him only a single time:

'I happened to be Dean of the Faculty of Economics and was invited to a dinner with Churchill before the conferring of a degree. During the dinner, I could see him swilling brandy in great quantities; by the time I was introduced to him, he could hardly speak but at once identified me as the author of The Road to Serfdom. *He was stock drunk. He said just one sentence: "You are completely right; but it will never happen in Britain". Half an hour later he made one of the most brilliant speeches I ever heard.'*[11]

Hayek agreed that it was quite possible that the 'Gestapo speech' cost Churchill the 1945 election. Whatever the reasons the political pendulum had now swung dramatically to the left in the UK.

Landslide to Labour under Clement Attlee

The new socialist government rapidly introduced its welfare state, drawing heavily on Beveridge's report on *Social Insurance and Allied Services*, and his *Full Employment in a Free Society*. Another of its priorities was large-scale nationalisation of key industries, which took place between 1945 and 1951, including the Bank of England, coal, civil aviation, telecommunications, transport, electricity, gas and iron and steel – industries covering millions of workers.

Back in London from the LSE evacuation to Cambridge, Hayek found himself increasingly isolated in his ideas. Keynes's *General Theory* now provided the postwar orthodoxy. So he busied himself in academic life, branching into new fields: political science, philosophy, sociology, psychology and the history of ideas.

FINAL NIGHT EXTRA

Evening Standard

ONE PENNY

In the family tradition BIRD'S CUSTARD

Britain swings to the Left—and the Churchill Government goes out in a landslide

SOCIALISTS IN

26 Ministers go down

LONDON, BIRMINGHAM CAPTURED: LIBERALS ECLIPSED

Britain has elected a Socialist Government. At 1.30 this afternoon the Conservative Party Central Office officially conceded the election to the Socialists. Latest position: For the Government 178; Against 355.

Three hours after the first General Election result was announced the Socialist lead was so commanding that there was no longer any doubt, and by half-past one they had won half the total number of seats, with 170 to come.

From the start the strong swing towards Socialism was clear. Minister after Minister fell—all three Service Ministers among them. So far 26 Ministers of the Caretaker Government have been defeated, five of them in the Cabinet.

The Socialists had five gains in the first half-hour. After an hour and three-quarters they had gained 55 seats and lost only two. By 2.30 their gains were 88 and losses two; Government gains were one, losses 82.

STATE OF PARTIES

FOR THE GOVERNMENT - 178
OTHER PARTIES - 361
SEATS TO BE DECLARED - 101

MINISTERS DEFEATED

MR. EDEN'S BIG MAJORITY

BEVIN AND MORRISON IN

NEW PRIME MINISTER

The next step

The Socialists win London

Evening Standard Reporter

The Socialists won London. They snatched 23 seats which were held by the Government, making a total of 48 seats out of a possible 62.

The Wiltshire anthem

Left hospital to fight his campaign

Election results in full are on Pages Three, Four, Five and Eight

In 1945, Hayek was invited by the new Director of the LSE, Sir Alexander Carr-Saunders, to write the 50th anniversary history of the London School of Economics. It was published as a 31-page article in *Economica* in February 1946. After noting that no fewer than four of its former teachers were in the present government (including Prime Minister Attlee and Chancellor of the Exchequer Hugh Dalton), Hayek concluded: '*Today, as before, the School harbours representatives of all the major political groups as well as probably an even larger number of men and women who, remote from all political connections, are devoted entirely to the advancement of their chosen subjects.*' Not necessarily by choice, perhaps, Hayek seems to have included himself among the latter.

In 1948 Hayek's next major book at the LSE was published, under the title *Individualism and Economic Order*. It contains a number of his essays on the problems of socialist calculation, exploring the various ways (including the use of prices and competition) in which socialist states attempt to solve the difficulties of allocating resources efficiently. The same book contains other essays on the nature of individualist philosophy and the strategy of the social sciences.

Hayek's contribution to the LSE was huge. Richard Cockett, a historian of the classical liberal revival in Britain, has observed that: 'The relationship between Hayek and Robbins was to become crucial to the development of a school of economic liberalism in Britain during the 1930s and 1940s, just as their work at the LSE was to make its economics department the outstanding centre of economic liberalism in Europe.'[12]

But following the Labour landslide in 1945, Hayek's ideas looked set to be sidelined as Keynesians and Fabians consolidated their power.

The weekly magazine *Illustrated* compiled a photo-feature on the LSE in 1948. 'Spellbound class listens to Professor F A Hayek, Tooke Professor of Economic Science and Statistics and author of anti-Socialist book *The Road to Serfdom*', is how *Illustrated* captioned this one.

Praising Hayek's impact at the LSE, Robbins wrote of his friend and colleague that: 'His erudition was immense: however much certain of his views have been attacked, no one to my knowledge has ever questioned his standing as an international scholar. As a leading member of the so-called younger Austrian School of the period, the contemporary of Gottfried Haberler, Fritz Machlup and Oskar Morgenstern, he brought to the School a host of intellectual contacts from abroad which otherwise would not have existed, and these not only in the sphere of technical economics: it is worth noting *en passant* that we owe the presence of Karl Popper, one of the most important appointments ever made at the School, to his introduction.'[13]

Illustrated's caption reads: 'Senior members of teaching staff converse in Founder's Room beneath portrait of Lord and Lady Passfield (Sidney and Beatrice Webb). Left to right: Dr G Willoughby, Professor F A Hayek, Mrs J Hood and Professor M Ginsberg'.

Karl Popper and Friedrich Hayek

Hayek had come across Popper's work in 1935 when Popper was teaching at the University of Vienna. Hayek found Popper's book on scientific method, *Logik der Forschung* (*The Logic of Scientific Discovery*), helpful in shaping his approach to empiricism and in his own attack on the assumptions of positivist economics.[14]

Popper was 'greatly encouraged' when Hayek invited him to read a paper on the book at his seminar at the LSE in October 1935 – the start of a 60-year friendship.

Popper left the University of Vienna in 1937 as the Nazis advanced across Europe, and went to Christchurch, New Zealand. There, as a Senior Lecturer in Philosophy at Canterbury University College, he wrote *The Open Society and its Enemies*, published in 1945.

In the two-volume study, which with *The Road to Serfdom* was to become one of the most important books in the defence of the free society, Popper attacked those he saw as the greatest enemies of democracy: Plato, Marx and Hegel. In his Acknowledgements, Popper wrote: 'I am deeply indebted to Professor F A von Hayek. Without his interest and support this book would not have been published.'

In 1945 Popper joined the staff of the LSE as Reader in Logic and Scientific Method, and was appointed Professor in 1949, holding the post until his retirement in 1969.[15] Arguably one of the most brilliant minds of the 20th century, he and Hayek remained lifelong friends.

By the late 1940s, it was obvious to Hayek that the need to reappraise the principles of the liberal society in the aftermath of the Second World War had never been greater. His fertile mind was to come up with a simple but, as it turned out, an extremely effective solution.

Karl (later Sir Karl) Popper (1902–1994).

Popper's own annotated copy of *Logik der Forschung*.

7 The rebirth of a liberal movement in Europe

Hayek founds the Mont Pèlerin Society, 1947

For liberal intellectuals, the world in the late 1940s, particularly in Europe, continued to be a hostile place. After the defeat of Hitler, an 'Iron Curtain' had descended across Europe, and communism now controlled the lives of millions of people. In Western Europe, many countries' populations looked to their socialist governments, such as that elected in Britain in the landslide of 1945, for the state planning that would 'win the peace' as it had won the war.

In Cambridge, Hayek was anxious that liberal principles should be revived, particularly in Germany, after the war was over. At a meeting in 1944 of the King's College Political Society in Cambridge he therefore proposed the formation of an international society of like-minded scholars *'to work out the principles which would secure the preservation of a free society'*.

Hayek's inspired proposal for an international society of liberal intellectuals came to practical fruition in 1947 on the shores of Lake Geneva in the Swiss Alps. Here, at Mont Pèlerin, 39 scholars from ten countries (including Ludwig von Mises and Lionel Robbins), all hand-picked and invited by Hayek, arrived to begin the immense task he had selected for them: *'the rebirth of a liberal movement in Europe'*.

Europe is divided by an 'Iron Curtain' stretching from the Baltic to the Adriatic.

'An iron curtain has descended across Europe'

The Daily Telegraph

LONDON, THURSDAY, APRIL 26, 1945

BERLIN IS COMPLETELY ENCIRCLED

RUSSIANS OVER ELBE NEAR AMERICANS

R.A.F. SMASH HITLER'S BERCHTESGADEN H.Q.

ALLIES CLOSE ON MUNICH DRIVE INTO REDOUBT

12,000-POUNDERS HIT FUEHRER'S CHALET

MOUNTAIN RETREAT BATTERED

ALLIES RACING ACROSS PLAINS OF LOMBARDY

TWO GENERALS CAPTURED

Mr. TRUMA[N] TO SAN [FRANCISCO]

"NOT ISOLA[TIONISM]" DARE NOT

4 FOREIGN M[INISTERS]

FORTRESSES HIT SKODA ARSENAL

GEN. PATTON

Daily Mail

WEDNESDAY, MAY 2, 1945

LATE WAR NEWS

HITLER DEAD, GERMAN RADIO TELLS WORLD

Admiral Doenitz is new Führer: 'The battle goes on'

ADOLF HITLER is dead. Grand Admiral Doenitz, Commander-in-Chief of the German Navy, has been appointed his successor. The German radio gave this news to the world at 10.25 last night in the following words: "It is reported from the Führer's headquarters that our Führer, Adolf Hitler, has fallen this afternoon in his command post in the Reich Chancellery fighting to his last breath against Bolshevism."

V DAY The full official plans

Nazi radio kept world in suspense

NO NEWS FROM HIMMLER, BUT

Surrender begins on three fronts

REPORTS received in London late last night indicated that, while no fresh offer of capitulation has been received from Himmler, large German forces on widely separated fronts have begun to surrender piecemeal to the Allies with or without his authority.

Busmen to strike to-day

FOOD SHIPS FOR DUTCH

After the six years it had taken to defeat Hitler, communist aggression threatened to replace the menace of National Socialism.

On 5 March 1946 Winston Churchill, Britain's wartime leader, gave his famous 'Iron Curtain' speech at Westminster College, Fulton, Missouri. Alarmed at the course of events since the Second World War, he told his audience:

'From Stettin in the Baltic to Trieste in the Adriatic, an iron curtain has descended across the continent. Behind that line lie all the capitals of the ancient states of Central and Eastern Europe. Warsaw, Berlin, Prague, Vienna, Budapest, Belgrade, Bucharest, and Sofia, all these famous cities and the populations around them lie in what I must call the Soviet sphere, and all are subject in one form or another, not only to Soviet influence but to a very high and, in many cases, increasing measure, of control from Moscow.'

In Walter Lippmann's dramatic phrase, the 'Cold War' had begun. But while Churchill was concerned about the developing military situation between the West and the East, Hayek had become increasingly anxious about the battle of ideas.

The Red Flag flies over the Reichstag in Berlin, 2 May 1945.

For Hayek, the intellectual threat was just as big as the military threat. After all, this was, in Max Hartwell's words, 'a world that had been devastated by two massively destructive wars and by a trade depression of unprecedented severity, a world whose traditions and liberal politics were threatened by the growth of governments and by increasing public intervention in economy and society.'[1] Hayek was determined not to lose that battle of ideas by default.

The minute book of the Political Society of King's College shows that, as early as 29 November 1943, he had discussed the question of *The Post-war Education of Germany*. Three months later, at a meeting of the same Political Society at King's College, on 28 February 1944 (a week before the publication of *The Road to Serfdom*), Hayek read a paper on *Historians and the Future of Europe*, concentrating particularly on the future of democracy in Germany in the aftermath of the war that was still raging, its outcome still uncertain.

It was at this meeting, under the chairmanship of the distinguished economic historian Sir John Clapham, that Hayek made his explicit proposal to form an international society of like-minded, liberal scholars which he thought *'could become a great force for good'* after the war.[2]

Hayek put much emphasis on *'facilitating contacts between individuals'* noting that: *'After the publication of* The Road to Serfdom, *I was invited to give many lectures. During my travels in Europe as well as in the United States, nearly everywhere I went I met someone who told me that he fully agreed with me, but that at the same time he felt totally isolated in his views and had nobody with whom he could even talk about them. This gave me the idea of bringing these people, each of whom was living in great solitude, together in one place. And by a stroke of good luck I was able to raise the money to accomplish this.'*[3]

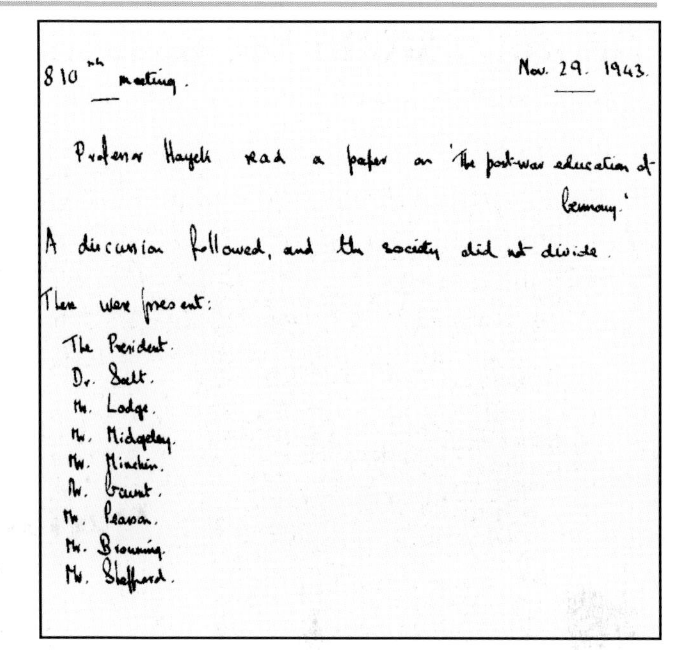

810th meeting. Nov. 29. 1943.

Professor Hayek read a paper on 'The post-war education of Germany.'

A discussion followed, and the society did not divide.

There were present:

The President.
Dr. Salt.
Mr. Lodge.
Mr. Midgeley.
Mr. Minchin.
Mr. Grant.
Mr. Pearson.
Mr. Browning.
Mr. Stafford.

*Historians and the Future of Europe**

Whether we shall be able to rebuild something like a common European civilization after this war will be decided mainly by what happens in the years immediately following it. It is possible that the events that will accompany the collapse of Germany will cause such destruction as to remove the whole of Central Europe for generations or perhaps permanently from the orbit of European civilization. It seems unlikely that, if this happens, the developments can be confined to Central Europe; and if the fate of Europe should be to relapse into barbarism, though ultimately a new civilization may emerge from it, it is not likely that this country would escape the consequences. The future of England is tied up with the future of Europe, and, whether we like it or not, the future of Europe will be largely decided by what will happen in Germany. Our efforts at least must be directed towards regaining Germany for those values on which European civilization was built and which alone can form the basis from which we can move towards the realization of the ideals which guide us.

Before we consider what we can do to that end, we must try to form a realistic picture of the kind of intellectual and moral situation we must expect to find in a defeated Germany. If anything is certain it is that even after victory we shall not have it in our power to make the defeated think just as we would wish them to; that we shall not be able to do more than assist any promising development; and that any clumsy efforts to proselytize may well produce results opposite to those at which we aim.

* A paper read to The Political Society at King's College, Cambridge, on February 28, 1944. The chair was taken by Sir John Clapham. Not published before.

In April 1947 Hayek's ambitious idea for an '*international academy of political philosophy*' came to fruition in the Swiss Alps at the Hotel du Parc in the small town of Mont Pèlerin, above Vevey on the shores of Lake Geneva.

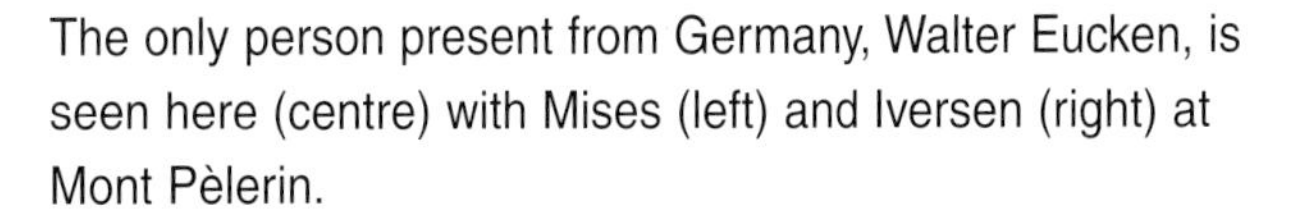

The only person present from Germany, Walter Eucken, is seen here (centre) with Mises (left) and Iversen (right) at Mont Pèlerin.

Wilhelm Röpke, now in Switzerland and soon to play a key role in Germany's post-war economic revival, was also invited to Mont Pèlerin.

A total of 39 scholars from ten countries arrived to begin the task outlined by Hayek – nothing less than replacing the post-war interventionist orthodoxy. Most of the Austrian School economists were already teaching in the United States, including Mises, Machlup and Karl Brandt. Yet even in Germany a spark of the liberal tradition remained, represented on this occasion by Hayek's old friend and leader of the liberal 'Freiburg School' of economics, Walter Eucken.

There is little doubt about the weight of intellect that Hayek assembled at the first meeting. As his copy of the mimeographed list of the participants shows, they came from the leading centres of liberal thought in the immediate postwar world. Principal among them were scholars from the LSE, Manchester, the Universities of Vienna, Freiburg and Chicago (it was the young Milton Friedman's first visit to Europe), as well as institutions in France, Switzerland and Norway.

Mont Pèlerin Conference.
1st - 10th April 1947.

List of Participants.

Prof.M.Allais, Ecole Nationale Supérieure des Mines, 218, Boulevard St.Germain, PARIS 7e

Prof.C.Antoni, Istituto Nazionale per le Relazioni Culturali con l'Estero, Piazza Firenze 27, ROME

Prof.K.Brandt, Food Research Institute, Stanford University, CALIFORNIA, U.S.A.

Prof.H.Barth, Université de Zurich, Heilighüsli 18, ZURICH.

Mr.J.Davenport, Fortune Magazine, 350 Fifth Ave. NEW YORK.

Prof.S.Dennison, Gonville & Caius College, CAMBRIDGE

Prof.A.Director, The Law School, University of Chicago, CHICAGO 37, ILLINOIS, U.S.A.

Prof.W.Eucken, Goethestrasse 10, FREIBURG i.Br.

Dr.E.Eyck, Chilswell House, Boars Hill, nr.OXFORD

Prof.M.Friedman, Dept.of Economics, University of Chicago, CHICAGO 37, ILLINOIS, U.S.A.

H.D.Gideonse, President, Brooklyn College, Bedford Ave, & Avenue H., BROOKLYN 10, N.Y., U.S.A.

Prof.F.D.Graham, Princeton University, 214, Western ay PRINCETON, New Jersey, U.S.A.

Prof.F.A.Harper, The Foundation for Economic Education, IRVINGTON-on-HUDSON, N.Y., U.S.A.

Prof.F.A.Hayek, The London School of Economics & Political Science, Houghton St. Aldwych, LONDON, W.C.2.

Mr.H.Hazlitt, Newsweek, Newsweek-Bldg, Broadway & 42nd St., New York 18, N.Y., U.S.A.

Dr.T.J.B.Hoff, Roald Amundsensgate, 1, OSLO.

Dr.A.Hunold, Bühnishof, Feldmeilen, ZURICH.

Mr.B.de Jouvenel, CH

-2-

Prof.C.Iversen, Ny Toldsbotgade 49, COPENHAGEN

Prof.J.Jewkes, Dept.of Economics, The University of Manchester, MANCHESTER.

Prof.F.H.Knight, University of Chicago, CHICAGO 37, Illinois, U.S.A.

Prof.F.Machlup, University of Buffalo, N.Y., U.S.A.

+Mr.L.B.Miller, Director of Bureau of Governmental Research, 153 E.Elisabeth, DETROIT 1, Michigan, U.S.A.

Prof.L.v.Mises, 777 West End Ave, New York 25, U.S.A.

Mr.Felix Morley, 1 Wetherill Rd., Washington 16, D.C. U.S.A.

+ Prof.M.Polanyi, The University of Manchester, Dept. of Chemistry, MANCHESTER.

Dr.K.R.Popper, The London School of Economics, LONDON, W.C.2.

+Prof.W.E.Rappard, Institut Universitaire des Hautes Etudes Internationales, 132 rue de Lausanne, GENEVE.

Mr.L.E.Read, President, The Foundation for Economic Education, IRVINGTON-on-HUDSON, N.Y., U.S.A.

Prof.L.Robbins, The London School of Economics & Pol. Science, Houghton St., Aldwych, LONDON, W.C.2.

+ Prof.W.Röpke, Institut Universitaire des Hautes Etudes Internationales, 45 Ave.de Champel, GENEVE.

Prof.G.J.Stigler, Brown University, PROVIDENCE R.I. U.S.A.

Prof.H.Tingsten, Dagens Nyheter, STOCKHOLM.

Prof.F.Trevoux, 10 Rue Duquesne, LYON.

Mr.V.O.Watts, The Foundation of Economic Education, IRVINGTON-on-HUDSON, New York, U.S.A.

Miss C.V.Wedgwood, Time & Tide, 32 Bloomsbury St., LONDON, W.C.1.

-3-

List of Observers.

Mr.H.C.Cornuelle, The Foundation for Economic Education, IRVINGTON-on-HUDSON, N.Y., U.S.A.

Mr.H. de Lovinfosse, Château Roos, WAASMUNSTER, Belgique

Mr.G.Révay, The Readers' Digest, 216 Bd.Saint Germain, PARIS.

Hayek's genius was to bring all these people together in one place. Thanks to eight years of wartime travel restrictions and then exchange controls, many had never met each other before coming to Mont Pèlerin.

In addition to the 20 professional economists, there were historians, lawyers and political philosophers, all in accordance with Hayek's multidisciplinary approach. There were also leading journalists, including Henry Hazlitt, formerly of the *New York Times* and then of *Newsweek*. Hazlitt, who had reviewed *The Road to Serfdom* in 1944, provided one of the vital communications links with the general public in America. The liberal fightback had begun.

Dr Albert Hunold (1899–1981), a Swiss businessman, seen here with Lionel Robbins, was responsible for the fund-raising in Europe for the conference and for all its organisation. The Volker Fund in the United States also provided substantial financing for the participation of the American scholars and Leonard Read's Foundation for Economic Education in New York helped with the detailed arrangements.

One of the informal photographs of the conference taken by Dr Albert Hunold.
From the left: Friedman, Robbins, Wedgwood, Director, Machlup, Röpke, Miss Salter (the secretary), Rappard, Mises, Eucken, Stigler, Davenport, Iversen.

Hayek's address to the conference

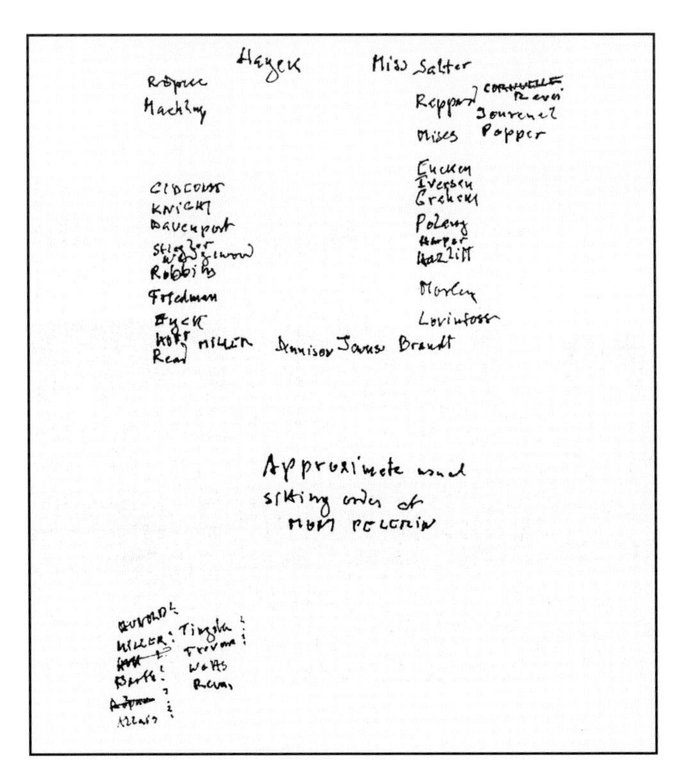

Hayek's own hand-written seating plan for the conference.

Hayek opened the conference, mindful of how much he was asking of the participants. '*I must confess that now, when the moment has arrived to which I have looked forward so long, my feeling of intense gratitude to all of you is much tempered by an acute sense of astonishment at my temerity in setting all this in motion, and of alarm about the responsibility I have assumed in asking you to give up so much of your time and energy to what you might well have regarded as a wild experiment....*

'*The basic conviction which has guided me in my efforts is that, if the ideals which I believe unite us, and for which, in spite of so much abuse of the term, there is still no better name than liberal, are to have any chance of revival, a great intellectual task must be performed....*'

Turning to the conference programme he had laid out, Hayek said: '*Of the subjects which I have suggested for systematic examination... the first is the relation between what is called "free enterprise" and a really competitive order.... I have taken the liberty to ask Professor Aaron Director of Chicago, Professor Walter Eucken of Freiburg and Professor Maurice Allais of Paris to introduce the debate on this subject....*

'*You will probably agree that the interpretation and teaching of history has during the past two generations been one of the main instruments through which essentially anti-liberal conceptions of human affairs have spread.... I am very glad that Miss Wedgwood and Professor Antoni have consented to open the discussion on this question....*

'*...two further topics... the problem of the future of Germany, and that of the possibilities and prospects of a European federation, seemed to me problems of such immediate urgency that no international group of students of politics should meet without considering them....*'[4]

Mont Pèlerin Conference
April 1st – 10th 1947.

PROCEEDINGS

Tuesday, April 1st. — Morning. Chair: Rappard
Report on aims and organisation of Conference by Hayek, Discussion of programme and time table. Allais, Dennison, Friedman, Hayek, Hunold elected as Standing Committee, Hayek and Hunold to act as Conference Secretaries.

Afternoon & Evening. Chair: Rappard
"Free" Enterprise or Competitive Order
Discussion opened by Hayek, Director & Eucken.
Speakers: Iversen, Eucken, Miller, Mises, Robbins, Polanyi, Friedman, Hazlitt, Rappard, Jouvenel, Gideonse, Tingsten.

Wednesday, April 2nd — Morning. Chair: Robbins
Modern Historiography & Political Education
Discussion opened by Wedgwood & Antoni.
Speakers: Eyck, Tingsten, Hayek, Popper, Mises, Rappard, Roepke, Barth Knight.

Afternoon & Evening. Chair: Iversen
The Future of Germany
Speakers: Eucken, Brandt, Robbins, Friedman, Rappard, Graham, Stigler.

Thursday, April 3rd — Morning & Evening. Chair: Allais
The Problems & Chances of European Federation
Discussion opened by Jouvenel
Speakers: Lovinfosse, Iversen, Morley, Rappard, Knight, Popper, Polanyi, Trevoux, Mises, Hayek, Machlup, Graham

The afternoon was occupied by an excursion to Chateau de Coppet.

-2-

Friday April 4th — Morning. Chair: Eucken
Liberalism & Christianity
Discussion opened by Knight
Speakers: Hoff, Davenport, Allais, Hayek, Polanyi, Popper, Gideonse, Roepke, Brandt, Jouvenel, Graham, Morley.

Afternoon. Chair: Antoni
General Discussion of aims and purposes of a permanent organisation.
Speakers: Hayek, Hazlitt, Robbins, Knight, Stigler, Machlup, Jewkes, Graham, Hoff, Jouvenel, Friedman, Popper, Polanyi, Dennison.
Committee appointed to draft statement of aims consisting of Eucken, Gideonse, Hayek, Hazlitt, Iversen, Jewkes.

Saturday, Apl.5th & Sunday Apl.6th — Excursion to Schwyz and Einsiedeln.

Monday April 7th. — Afternoon. Chair: Hoff
Contra-Cyclical Measures, Full Employment & Monetary Reform.
Opened by Stigler
Speakers: Graham, Hazlitt, Robbins, Director, Roepke, Allais, Knight, Jouvenel, & Friedman

Evening
Discussion of Draft Statement of Aims

Tuesday April 8th — Morning. Chair: Graham
Wage Policy & Trade Unions
Opened by Machlup
Speakers: Allais, Lovinfosse, Graham, Dennison, Polanyi, Iversen, Knight, Rappard, Watts, Brandt, Jouvenel, Davenport, Jewkes.

Evening. Chair: Mises
Taxation, Poverty & Income Distribution
Opened by Friedman
Speakers: Jewkes, Miller, Polanyi, Dennison, Rappard, Allais, Jouvenel, Hayek, Knight, Brandt, Popper, Watts.

-3-

Wednesday April 9th — Morning. Chair: Read
Agricultural Policy
Opened by Brandt
Speakers: Director, Roepke, Miller, Robbins, Hayek, Eucken, Rappard, Graham, Mises.

Afternoon. Chair: Morley
Discussion of Organisation of permanent body.

Evening. Chair: Gideonse
The Present Political Crisis
Opened by Polanyi
Speakers: Davenport, Jouvenel, Knight, Popper, Brandt, Watts, Wedgwood, Robbins.

Thursday April 10th — Morning. Chair: Hayek
Discussion and adoption of "Memorandum of Association" of Mont Pèlerin Society.

Hayek's own copy of the mineographed programme. Topics ranged from *'Free' Enterprise or Competitive Order* to *Modern Historiography & Political Education*, and from *Liberalism & Christianity* to *Wage Policy & Trade Unions*.

After Hayek had delivered his opening remarks, the participants began a week's discussions. However, it was not all work. During the conference the participants went to the Chateau de Coppet, Schwyz and Einsiedeln. So began a delightful tradition that each host country has sought to emulate at subsequent meetings of the Mont Pèlerin Society around the globe.

Dr Hunold's informal photographs of the participants at the conference, reproduced from Hayek's own bound presentation album.

(l to r) Iversen, Graham, Polyani, Hazlitt, Morley, Lovinfosse, Brandt.

(l to r) Rappard, Popper, Mises.

(l to r) Jewkes, Read, Eyck, Röpke, Friedman, Robbins, Wedgwood.

Chicago Tribune
3 April 1947

7 NATIONS MAP FREEDOM FIGHT IN SECRET TALK

Hope to Draft Plan of General Appeal

BY E. R. NODERER
[Chicago Tribune Press Service]

MONT PELERIN, Switzerland, April 2—Forty-two economists and publicists from seven countries are meeting privately here to discuss plans for a permanent organization to study the "philosophy of freedom." Friedrich A. Hayek of the London School of Economic and Political Science, who arranged the conference and issued the invitations, said an announcement "might" be made after the meeting ends April 10.

The question of publicity was taken up at the opening session yesterday. After an hour's debate, it was decided to bar the press and a six man committee appointed to draft a statement to hand to any inquiring reporters who might show up. Dr. Hayek said it would be carefully worded so as to contain no information.

Churchill Adviser Explains

Reasons for the secrecy were not altogether clear, tho Hayek, who is Winston Churchill's adviser on economic matters, said that since nothing may come of the conference he didn't feel publicity was desirable.

It was learned that Hayek defined the purpose of the conference as an attempt "to enlist the support of the best minds in formulating a program which has a chance of gaining general support. Our effort therefore differs from any political task in that it must be essentially a long run effort concerned not so much with what would be immediately practicable but with the belief which must gain ascendence if the dangers are to be averted which at the moment threaten individual freedom."

Topics of Discussion

Discussions have been scheduled on free enterprise or the competitive order, problems and chances of European federation, liberalism and Christianity, modern historiografy, and political education.

The countries represented are the United States, Britain, France, Italy, Switzerland, Norway, and Denmark. From the University of Chicago came Aaron Director, Milton Friedman, and Frank H. Knight.

The press takes an interest in the 'freedom fight' (*Chicago Tribune* 3 April 1947). Note that Dr Hayek is quoted as saying with aplomb that the press release 'would be carefully worded so as to contain no information'!

These photographs taken by Dr Hunold record an excursion enjoyed by the assembled scholars.

After a week of formal and informal discussions, all those present at Mont Pèlerin agreed that the conference had been very successful and the Society should continue. Despite Hayek's original suggestion that his 'Academy' should be known as the 'Acton–Tocqueville Society', the name 'Mont Pèlerin Society' was adopted (partly because agreement could not be reached, and partly because the participants felt that linking the name of the society to those of historical individuals would inevitably put limits on its future discussions). Looking back some 35 years later Hayek said he considered founding the Mont Pèlerin Society one of his most significant achievements:

'In some way, the founding and the first conference of the Mont Pèlerin Society, which, I feel entitled to say, was my own idea, although I received a great deal of support from Röpke as well as from Mises, constituted the rebirth of a liberal movement in Europe. Americans have done me the honour of considering the

'The rebirth of a liberal movement in Europe'

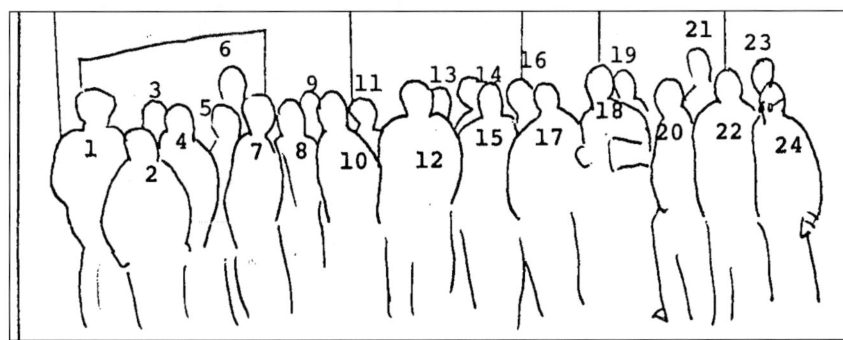

Hayek attended nearly all of the meetings, including this one held in 1972 at Montreux, near Mont Pèlerin, to celebrate the Society's twenty-fifth anniversary.

1. Trygve J B Hoff (Norway)
2. Milton Friedman (USA)
3. Stanley Dennison (UK)
4. Henry Hazlitt (USA)
5. Hans Ilau (Germany)
6. Max Thurn (Austria)
7. F A Lutz (Switzerland)
8. Herbert Frankel (UK)
9. Friedrich Hayek (UK)
10. Nicomedes Zuloaga (Venezuela)
11. Arthur Shenfield (UK)
12. Christian Gandil (Denmark)
13. Ralph Harris (UK)
14. Ole-Jacob Hoff (Norway)
15. Jean-Pierre Hamilius (Luxembourg)
16. Benjamin A Rogge (USA)
17. Arthur Kemp (USA)
18. Leonard E Read (USA)
19. Gunther Schmolders (Germany)
20. Richard A Ware (USA)
21. George Stigler (USA)
22. Fritz Machlup (USA)
23. James Buchanan (USA)
24. Carl Iversen (Denmark)

There were others present at the meeting who were not captured on this archive photograph.

publication of The Road to Serfdom *as the decisive date, but it is my conviction that the really serious endeavour among intellectuals to bring about the rehabilitation of the idea of personal freedom especially in the economic realm dates from the founding of the Mont Pèlerin Society in 1947.'*[5]

Hayek served as President of the Society for 13 years until 1960, and then as its Honorary President from 1960 until he died in 1992. When the Society celebrated its 50th anniversary in 1997, it had held 28 general meetings and 19 regional meetings around the globe, from Vienna to Vancouver, Cambridge to Tokyo.

The power of ideas

Hayek understood the extent of the power of ideas and knew the challenges facing him and other liberal intellectuals in the postwar world. At the LSE in the late 1940s, and indeed in Britain generally, Hayek found his approach increasingly out of favour, particularly in academic circles.

Keynes in his library at 46 Gordon Square, London.

His old adversary, Keynes, had died in 1946, but so great was the extraordinary and pervasive impact of Keynesianism on Britain that the arguments of the economic liberals against it hardly received a public hearing.[6]

In the 1948 publication of his paper *Free Enterprise and Competitive Order*, given at the first Mont Pèlerin Conference, Hayek said: *'I do not find myself often agreeing with the late Lord Keynes, but he has never said a truer thing than when he wrote… that "the ideas of economists and political philosophers, both when they are right and when they are wrong, are more powerful than is commonly understood. Indeed the world is ruled by little else. Madmen in authority, who hear voices in the air, are distilling their frenzy from some academic scribbler of a few years back… soon or late, it is ideas, not vested interests, which are dangerous for good and evil".'*[7]

Not just Keynes, but the Webbs too, achieved this lasting influence over ideas and policy. Remembering his youthful dalliance with Fabian Socialism, Hayek's 1948 review in *Economica* in August 1948 of *Our Partnership* by Beatrice Webb repeats his theme that ideas have consequences:

'The strongest impression left by this second part of Beatrice Webb's memoirs is that she and her husband owed the extent of their influence largely to the fact that they cared only for the success of the ideas in which they believed, without any regard to who got the credit for them, that they were willing to operate through any medium, person or party which allowed itself to be used, and above all, that they fully understood, and knew how to make use of, the decisive position which the intellectuals occupy in shaping public opinion.'[8]

As Hayek was writing this, the Labour government of Clement Attlee was busy nationalizing the 'commanding heights' of the economy and creating the post-war welfare state. Hayek was in no doubt that the Webbs were part of that *'small group of people whose ideas have changed Great Britain in the past forty years and rule it at present'*.

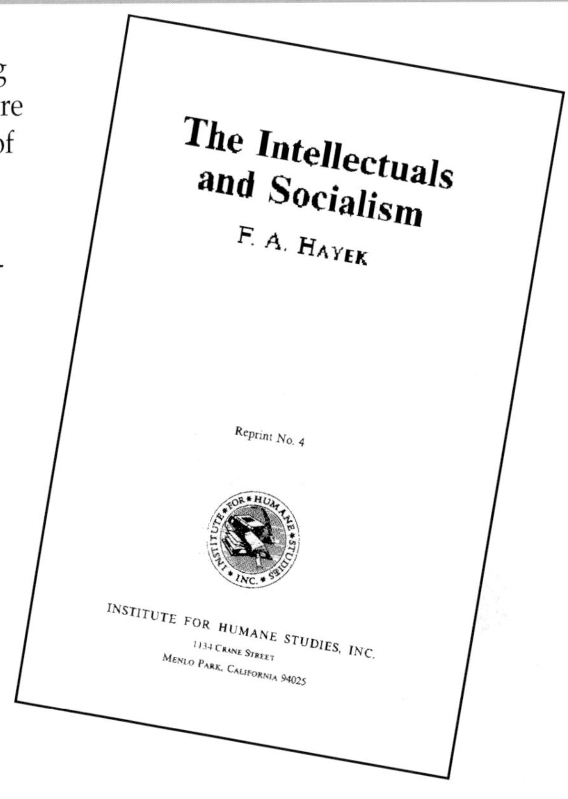

The Intellectuals and Socialism

F. A. HAYEK

Reprint No. 4

INSTITUTE FOR HUMANE STUDIES, INC.
1134 CRANE STREET
MENLO PARK, CALIFORNIA 94025

Hayek developed his arguments about the role of intellectuals in influencing public policy in his famous, often reprinted article, *The Intellectuals and Socialism*, published in 1949:

'The main lesson which the true liberal must learn from the success of the socialists is that it was their courage to be Utopian which gained them the support of the intellectuals and therefore an influence on public opinion which is daily making possible what only recently seemed utterly remote.... Unless we can make the philosophic foundations of a free society once more a living intellectual issue, and its implementation a task which challenges the ingenuity and imagination of our liveliest minds, the prospects of freedom are indeed dark. But if we can regain that belief in the power of ideas which was the mark of liberalism at its best, the battle is not lost. The intellectual revival of liberalism is already under way in many parts of the world. Will it be in time?'[9]

Hayek in 1950 at the meeting of the Mont Pèlerin Society at Bloemendaal, Holland.

Within a year of writing these words, Hayek's career was to lead him back to another of those parts of the world, the United States. His decision to take up an attractive teaching post at the University of Chicago in 1950 meant that Britain lost its most prominent liberal intellectual; but America was to gain a powerful new focus for the liberal revival there.[10]

8 At the University of Chicago, 1950–1962

Hayek's appointment to the Committee on Social Thought

In 1950, Hayek decided to leave the LSE when he was invited to the University of Chicago. He was appointed a Professor not of Economics but of Social and Moral Science and became a member of the influential Committee on Social Thought where a galaxy of extraordinary minds had coalesced.

For the next 12 years, Hayek's talents blossomed in the buoyant, free enterprise society of postwar America. He became a true polymath, shifting his focus from pure economics to the broader philosophical questions underpinning the future of Western civilisation.

Hayek's international contacts were maintained at regular meetings of the Mont Pèlerin Society, held in a variety of countries, and through his Alpine retreats and summer schools in Austria.

He remarried in 1950 after a divorce from Hella that would sadly estrange him from many of his friends in Europe, including Lionel Robbins, his early colleague at the LSE.

Yet the Austrian Tyrol continued to offer him the inspiration and solitude to write, and it was there that he worked on *The Constitution of Liberty*. Published simultaneously in 1960 in both America and England, the book was hailed as a masterpiece and his greatest work in political philosophy.

By the early 1960s, Hayek was recovering from a severe depression. Family roots and his love of the Austrian Alps were now to bring him back to Europe.

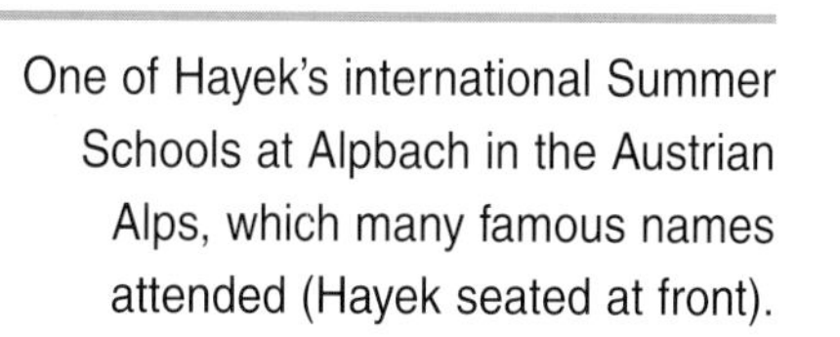

One of Hayek's international Summer Schools at Alpbach in the Austrian Alps, which many famous names attended (Hayek seated at front).

Hayek moves to the United States

By the late 1940s, settled back with the LSE in London and following the successful establishment of the Mont Pèlerin Society, Hayek was encouraged by his various lecturing visits to the United States to make a dramatic change in his career. In 1950, for professional and personal reasons (including his divorce in England and marriage to Helene Bitterlich), Hayek accepted an invitation to join the teaching staff of the University of Chicago, with financial support from the Volker Fund.

Hayek's divorce and remarriage had estranged him from family friends such as the Robbinses. However, there were other reasons for leaving too: *'I should never have wished to leave England, certainly not if I could have continued to live in Cambridge [even though] I did begin to feel the atmosphere of a highly specialised institution like LSE somewhat narrow and the work there – because of the long distances to travel and the evening teaching – increasingly exhausting.'*[1]

However, Hayek's ideas were iconoclastic even in America. So in the event, he ended up going to Chicago not as Professor of Economics but as Professor of Social and Moral Science and a member of the Committee on Social Thought. Hayek noted: *'Whoever was behind wanting me there was persistent and shifted his efforts from the Faculty of Economics to the Committee on Social Thought. Which to me ultimately was much more attractive, because after teaching economic theory for twenty years I was a little tired, and the invitation from the Committee on Social Thought indicated that I could lecture on any borderline subject in the social sciences and, if at any time I didn't want to teach, it would not be required. This with a correspondingly higher salary was irresistible. So I did go to Chicago.'*[2]

Hayek was already well known in the United States, where *The Road to Serfdom* had sold at an unprecedented rate for such a work – particularly since it had been written for a European readership. Significantly, the American publication of *The Road to Serfdom* had been undertaken by the University of Chicago, where at that time the famous 'Chicago School' of economics was flourishing. This School was led by Frank Knight, Milton Friedman and later by George Stigler, all Hayek's guests at the founding meeting of the Mont Pèlerin Society in 1947.

As Hayek reflected many years later: *'Chicago was a very exciting university'*.[3] Arthur Shenfield, President of the Mont Pèlerin Society 1972–1974, said of it: 'The University of Chicago was the most celebrated centre in the US of scholars championing the free market economy and the free society of which that economy is the shield and support.'[4] And it was under American leadership that postwar Europe was being rebuilt and communism contained.

Hayek at Chicago with some of his colleagues who were members of the Committee on Social Thought. Left to right: Otto von Simson, James Redfield, Peter van Blanckenhagen, John Nef, Frank Knight, Friedrich Hayek.

Hayek in his study at Chicago, with his comprehensive library....

Hayek had been invited to Chicago by Aaron Director and John Nef. Director had been influential in getting *The Road to Serfdom* published by the University of Chicago Press. Having founded law and economics at Chicago, he had been one of Hayek's guests at the first meeting of the Mont Pèlerin Society. Nef was an economic historian who founded the Committee on Social Thought and was its chairman 1945–1964.[5]

The Committee on Social Thought, which 'brought together different branches of the social sciences, became one of the most prestigious academic bodies in the United States'.[6] Hayek revelled in the new academic freedom it gave him.

'In fact the post at the Committee on Social Thought offered me almost ideal opportunities for the pursuit of the new interests I was gradually developing. As Professor of Social and Moral Science, I was allowed there to devote myself to almost any subject.... I had, as a matter of fact, become somewhat stale as an economist and felt much out of sympathy with the direction in which economics was developing. Though I had still regarded the work I had done during the 1940s on scientific method, the history of ideas, and political theory as temporary excursions into another field, I found it difficult to return to systematic teaching of economic theory and I felt it rather as a release that I was not forced to do so by my teaching duties.'[7]

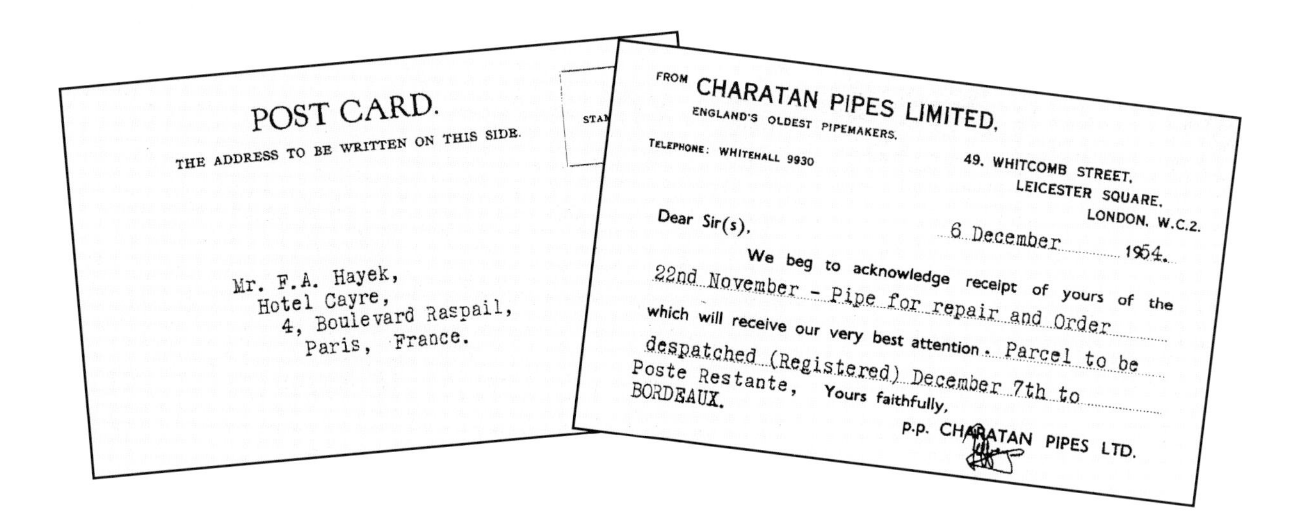

POST CARD.

THE ADDRESS TO BE WRITTEN ON THIS SIDE.

Mr. F.A. Hayek,
Hotel Cayre,
4, Boulevard Raspail,
Paris, France.

FROM CHARATAN PIPES LIMITED.
ENGLAND'S OLDEST PIPEMAKERS.

TELEPHONE: WHITEHALL 9930

49, WHITCOMB STREET,
LEICESTER SQUARE,
LONDON, W.C.2.

Dear Sir(s), 6 December 1954.

We beg to acknowledge receipt of yours of the 22nd November - Pipe for repair and Order which will receive our very best attention. Parcel to be despatched (Registered) December 7th to Poste Restante, BORDEAUX.

Yours faithfully,
p.p. CHARATAN PIPES LTD.

...and the repairs to his pipe are obviously well in hand!

Just as he had taken part in Mises's *Privatseminar* in Vienna and then helped run the Grand Seminar at LSE with Robbins, so Hayek now had his own seminar and it was a brilliant success. 'At Chicago, he was (by general agreement) well able to supply the range of expertise needed for the post, and his emphasis on stimulating the open discussion continued. At weekly seminars arranged by him, for example, some of the best minds of the University were able to meet, without barriers of age, status or academic discipline, to discuss the wide range of topics he proposed.'[8]

It was no soft option. In one notice for the seminar, Hayek says firmly: '*Students of the departments of Economics, Political Science, Sociology, History and Philosophy and others specially interested in the subject, will be admitted if they undertake to acquaint themselves thoroughly with the work and the ideas of one of the persons or groups listed below so as to be able to represent their views on each topic, as it arises in the discussion.*'

Shirley Robin Letwin, who studied under Hayek at Chicago and attended the 1951/52 seminar, graphically described the proceedings:

'Hayek presided over this remarkable company with a gentle rectitude that made his seminar an exercise in the liberal virtues. Every remark, however fatuous, no matter how obscure or young the speaker, was heard to the very end with a respect that the weaker members found maddening. The general subject was liberalism and no one was in any doubt about Hayek's convictions… [his] conduct as a colleague and teacher was entirely of a piece with this impeccably liberal colloquium.'[9]

The Anglo-American political philosopher Shirley Robin Letwin (1924–1993)

THE UNIVERSITY OF CHICAGO

THE COMMITTEE ON SOCIAL THOUGHT

April 1951

During the session 1951-52 I hope to conduct for the Committee on Social Thought on Wednesday evenings (8-10) a seminar on

THE LIBERAL TRADITION

(from J. Locke to L. v. Mises)

The intention is to examine in a systematic order the position
taken by the chief representatives of classical liberalism on the
main political and economic issues, with the various authors represented by individual members of the seminar. Students of the departments of Economics, Political Science, Sociology, History and
Philosophy, and others specially interested in the subject, will
be admitted if they undertake to acquaint themselves thoroughly
with the work and the ideas of one of the persons or groups listed
under I below so as to be able to represent their views on each
topic as it arises in the discussion.

As it is desirable that the members of the seminar should
familiarize themselves with the ideas of their respective authors
before the discussions begin, it is hoped that it will be possible
to assign at least the main authors before the summer. Students
wishing to take part in the seminar should write to me during the
next few weeks, and at any rate before the middle of May, indicating their preferences as regards authors and stating their special interests and command of languages. The main references will
in most instances be found under the appropriate heading in the
Encyclopaedia of the Social Sciences which should be consulted before the interview in which I hope to discuss the matter with each
prospective member of the seminar when I may also be able to give
further references. A list of the chief general works on the
history of liberalism and a provisional list of the main topics to
be considered is given under II and III below.

F. A. Hayek

Hayek's synopsis of the 1951/52 seminar.

Hayek's seminar on the liberal tradition from Locke to Mises

His mimeographed list of topics (with handwritten amendments) gives an idea of Hayek's wide range of interests.

COMMITTEE ON SOCIAL THOUGHT
(Session 1951/52)

SEMINAR ON

"THE LIBERAL TRADITION"

1. Systematic Order of Topics

I. INTELLECTUAL FREEDOM
- Belief in power of ideas
  - diversity and limitation of individual intellect
  - that truth will emerge from the interplay of differ
    intellects in free discussion: reason a soci
    process: belief in persuasion.
  - that nobody is competent authoritatively to decide
    knows best
  - that even error has to be respected
  - that the spreading of opinion is inevitably a grad
    process and that therefore all aims must be
- Postulates: Tolerance
  - Freedom of Thought and Conscience
    - of Speech and Assembly
    - of the Press
  - Absence of all Forms of Censorship (also pos
    theatre, cinema, sale of pictures and b
  - Academic Freedom (Teaching and Research)
- Limitations: Tolerance for the advocacy of intolerance?
  - "Seditious", pornographic etc. literature
  - Defence Secrets and other official secrets
- Question: How essential is agreement on some fundamenta
  decent society?

II. ATTITUDE TO RELIGION, MORALS AND TRADITION
(The "RATIONALISM" OF LIBERALISM)
- General attitude to organized religion: Church and State
- Reason of Conflicts between Liberalism and Religion (esp. Ro
  Catholic Church)
- Source and Sanctions of Moral Beliefs, Their Role and Import
- The pressure of public opinion: Social habits and Customs
- General attitude to irrational elements in social relations

III. EDUCATION
- Compulsory?
- Free?
- State?
- "Education for Citizenships" and Propaganda

IV. CONSTITUTIONAL ARRANGEMENTS
- Democracy
- Universal Suffrage
- Majority Rule and its Limitations
- Republicanism
- Centralism versus Federalism
- Local Self Government
- Division of Powers

V. THE LAW
- Rule of Law, Equality before the Law
- Rules vs. Discretion
- Law and Freedom ("Freedom under the Law")
- Criminal Law: Nulla Poena sine lege
- Powers of the Police: (civic liberties)

VI. THE RESPONSIBLE INDIVIDUAL
- Treatment of Children
  - Sick and Disabled
  - Insane
  - Old
  - Poor
  - "Inferior" Races

VII. ATTITUDE TO SMALLER GROUPS AND ORGANIZATIONS
- Monism versus Pluralism
- The Family, Marriage and Sexual Relations
- Voluntary Organization
- Organized Charity

VIII. ECONOMIC ORGANIZATION: PROPERTY AND CONTRACT
- Basic Position of Individual Property, Rules of Expropriation
- Rule of the Market
- Monopoly
- Corporations
- Trade Unions
- Public Utilities
- Factory Laws and similar regulations
- Licensing of Dangerous Trades and Professions
- Patents and Copyright ("Prohibition" of Alcohol and Narcotics.)

IX. EQUALITY
- Taxation
- Redistribution
- Inheritance
- Equality versus Mobility

X. OTHER POSITIVE STATE FUNCTIONS
- Defense (Militia or Standing Army)
- Sanitation and Health
- Communications (Roads, Harbours, Lighthouses)
- Weights and Measures
- Information

XI. NATIONALISM
- "Self-determination"
- Right to Secession
- Delimitation of Political Unit
- Rights of Minorities
- Linguistic Problems

XII. INTERNATIONAL
- War
- International Law
- Principle of Non-Intervention
- Free Trade
- Capital Movements
- Migration, Racial Mixture
- Colonies and Undeveloped Countries

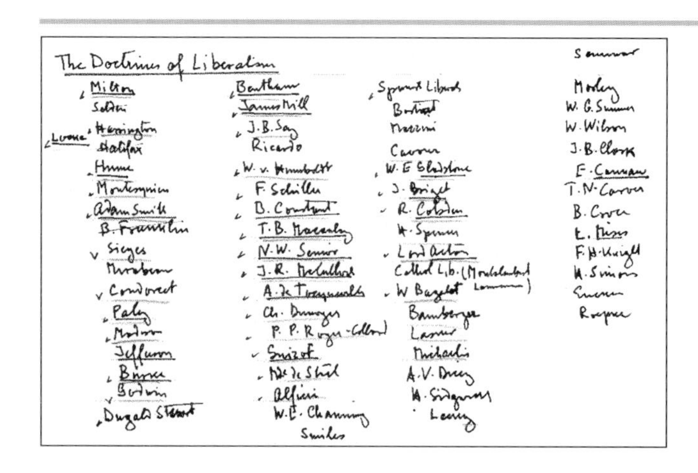

The Doctrines of Liberalism

Seminar

Milton, Selden, Harrington, Halifax, Hume, Montesquieu, Adam Smith, B. Franklin, Sieyes, Mirabeau, Condorcet, Paley, Madison, Jefferson, Burke, Godwin, Dugald Stewart

Bentham, James Mill, J. B. Say, Ricardo, W. v. Humboldt, F. Schiller, B. Constant, T. B. Macaulay, N. W. Senior, J. R. McCulloch, A. de Tocqueville, Ch. Dunoyer, P. P. Royer-Collard, Guizot, Alfieri, W. E. Channing, Smiles

Spanish Liberals, Bastiat, Mazzini, Cavour, W. E. Gladstone, J. Bright, R. Cobden, H. Spencer, Lord Acton, Bamberger, Lasker, Michaelis, A. V. Dicey, H. Sidgwick

Morley, W. G. Sumner, W. Wilson, J. B. Clark, E. Cannan, T. N. Carver, B. Croce, F. H. Knight, H. Simons, Roepke

Hayek's handwritten card index list of authors for the seminar.

II. Main Works on the History of Liberalism:

The best, but by no means altogether satisfactory general work is:

G. de Ruggiero, The History of European Liberalism, 1927.

This should be supplemented by some of the following:

F. M. Watkins, The Political Tradition of the West, 1948.
W. A. Orton, The Liberal Tradition, 1945.
J. Salwyn Shapiro, Liberalism and the Challenge of Fascism, the work and 1949.
L. T. Hobhouse, Liberalism, 1911.
H. Michel, L'idée de l'état, 1895.
H. J. Laski, The Rise of European Liberalism, 1936.
A. Schatz, L'Individualism economique et social, 1907.
F. Meinecke, Weltbürgertum und Nationalstaat, 1911.

and for the development in particular countries:

A. V. Dicey, Law and Public Opinion in England, 1914.
L. Diez del Corral, Il Liberalismo Doctrinalo, Madrid, 1945.
F. Federici, Der deutsche Liberalismus (tr. from the Italian) Zurich, 1941.
Klein-Hattingen, Geschichte des deutschen Liberalismus, 1911/12
O. Westphal, Welt- und Staatsauffassung des deutschen Liberalismus (Historische Bibliothek XLI) München, 1919.

Hayek's reading list of 'Main Works on the History of Liberalism' shows he expected his students to be familiar with French and German as well as English.

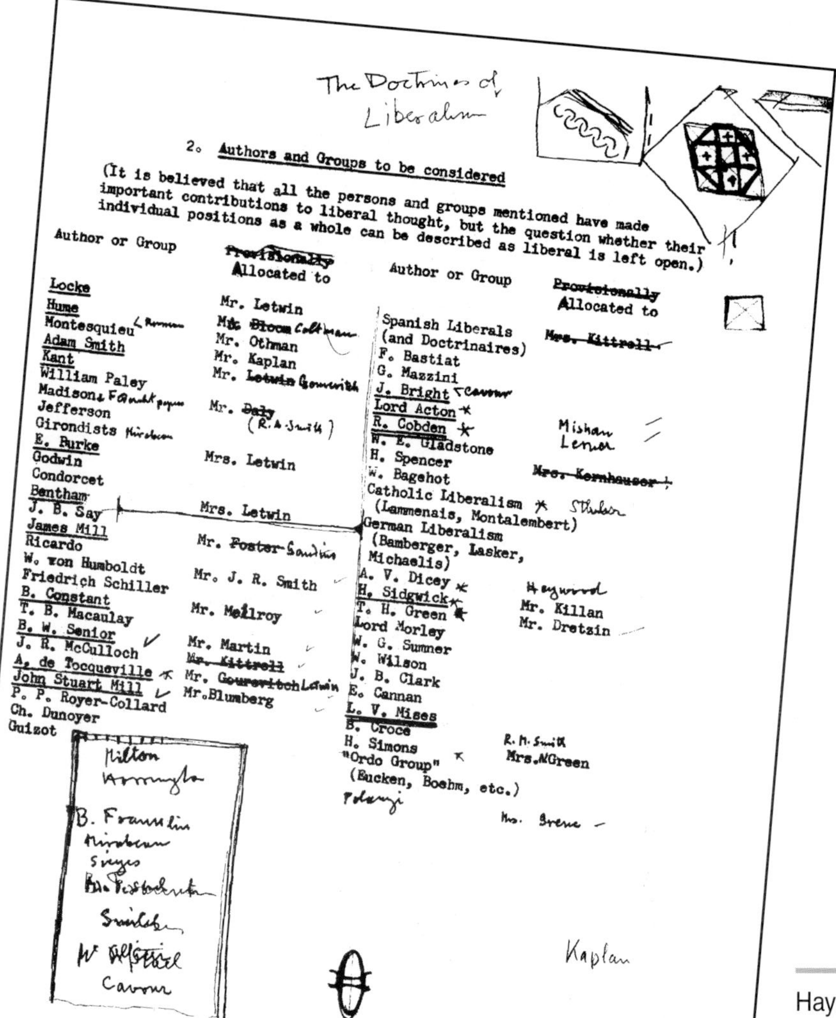

The Doctrines of Liberalism

2. Authors and Groups to be considered

(It is believed that all the persons and groups mentioned have made important contributions to liberal thought, but the question whether their individual positions as a whole can be described as liberal is left open.)

Author or Group — ~~Provisionally~~ Allocated to

Locke
Hume — Mr. Letwin
Montesquieu — Mr. ~~Bloom~~ Coltman
Adam Smith — Mr. Othman
Kant — Mr. Kaplan
William Paley — Mr. ~~Letwin~~ Gourevitch
Madison & Federalist papers
Jefferson — Mr. ~~Daly~~ (R. A. Smith)
Girondists Mirabeau
E. Burke
Godwin — Mrs. Letwin
Condorcet
Bentham
J. B. Say — Mrs. Letwin
James Mill
Ricardo — Mr. ~~Foster~~ Sanders
W. von Humboldt
Friedrich Schiller — Mr. J. R. Smith
B. Constant
T. B. Macaulay — Mr. McIlroy
B. W. Senior
J. R. McCulloch — Mr. Martin
A. de Tocqueville — ~~Mr. Kittrell~~
John Stuart Mill — Mr. ~~Gourevitch~~ Letwin
P. P. Royer-Collard — Mr. Blumberg
Ch. Dunoyer
Guizot

Author or Group — ~~Provisionally~~ Allocated to

Spanish Liberals (and Doctrinaires) — ~~Mrs. Kittrell~~
F. Bastiat
G. Mazzini
J. Bright — Cavour
Lord Acton
R. Cobden — Mishan
W. E. Gladstone — Lerner
H. Spencer
W. Bagehot — ~~Mrs. Kornhauser~~
Catholic Liberalism (Lammenais, Montalembert)
German Liberalism (Bamberger, Lasker, Michaelis)
A. V. Dicey
H. Sidgwick — Mr. Killan
T. H. Green — Mr. Dretzin
Lord Morley
W. G. Sumner
W. Wilson
J. B. Clark
E. Cannan
L. V. Mises
B. Croce
H. Simons — R. M. Smith
"Ordo Group" (Eucken, Boehm, etc.) — Mrs. Green
Polanyi

Milton
Harrington
B. Franklin
Mirabeau
Sieyes
Smiles
Cavour

Kaplan

Hayek's dittoed list of 'Authors and Groups to be considered' (and their allocation) complete with his doodles!

Hayek, now into his fifties, continued his prolific research and writing output. In 1951 he published *John Stuart Mill and Harriet Taylor: Their Friendship and Subsequent Marriage.*

Based on Mill's own papers lodged in England and America, the book had been a side-product of Hayek's other research. Although Hayek said later that Mill had never particularly appealed to him, he unintentionally became one of the foremost experts on the subject.

Thus the *Economist* said that 'Professor Hayek has arranged his material with great art and provided a commentary, admirable in brevity and taste, clearly founded on long research. The result is a most moving and fascinating book [which] tells us much more than Mill did himself about the years of his prime.'

The leading 19th-century British philosopher John Stuart Mill (1806–1873).

The Counter-Revolution of Science

STUDIES ON THE ABUSE OF REASON

By F. A. Hayek

The Free Press of Glencoe
Collier-Macmillan Limited, London

In 1952 Hayek published *The Counter-Revolution of Science: Studies on the Abuse of Reason*, a volume of essays in which he analyses the intellectual origins of engineering and discusses their disastrous political consequences, whether under socialism or fascism.

Now regarded as a classic in social science literature, the book explains with remarkable accuracy and detail the problems and mistakes that arise when we attempt to use the methods of the physical sciences in the study of society. For not only is society a complex phenomenon, says Hayek, and therefore quite unlike the simple models studied in the physical sciences, but each individual making up that complex structure is himself complex and impossible to predict with any accuracy.

In Hayek's view, 'The problem for any planner is that the "facts" he must deal with are not concrete things, but are the relationships and behaviour of individuals themselves, something which nobody can predict in advance. It is a poor basis for any social "science": while we might be able to talk about some general patterns of society, we should never suppose that we can completely predict it.'[10]

His next book, *The Sensory Order*, also published in 1952, expanded an idea in psychology that he had been developing since his days at the University of Vienna, where he was stimulated by the work of Ernst Mach, the Professor of the Inductive Sciences. Hayek later recalled that '*the insights I gained from the first stage in 1920 [and] later in the 1940s were probably the most exciting events that ever occurred to me, which shaped my thinking.*'[11]

The Sensory Order was, he wrote in the Preface, '*the outcome of an idea which suggested itself to me as a very young man when I was still uncertain whether to become an economist or a psychologist. But though my work has led me away from psychology, the basic idea then conceived has continued to occupy me; its outlines have gradually developed, and it has often proved helpful in dealing with the problems of the methods of the social sciences.*'

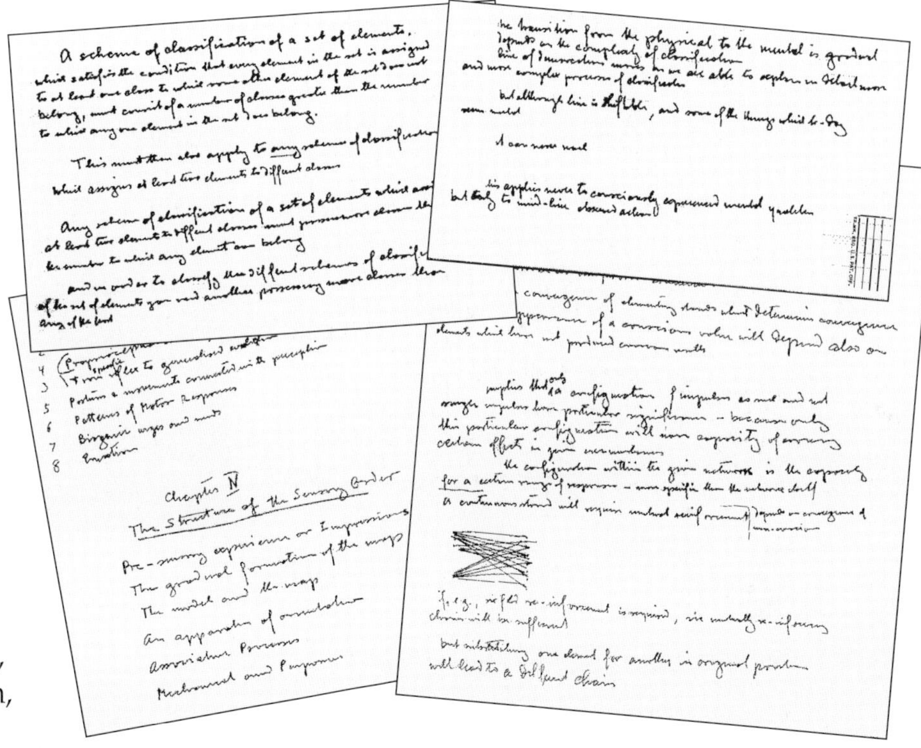
A scheme of classification of a set of elements.

Chapter IV

The Structure of the Sensory Order

Some of the notes for *The Sensory Order* that Hayek made over many years and took with him to Chicago.

Capitalism and the Historians, published in 1954 and edited, with an introduction, by Hayek, comprised essays by T S Ashton, Louis Hacker, W H Hutt and Bertrand de Jouvenel. The only publishing venture of the Mont Pèlerin Society, it arose from the main contentions of the original meeting in 1947: that historians had a significant power to influence socio-political values and attitudes. Drawing on papers presented to the meeting at Beauvallon in France,[12] it painstakingly refuted the widespread myth that early capitalism brought only poverty and misery to the downtrodden workers.

The book became 'part of the essential reading on the controversy about the effects of English industrialisation on the mass of the population.... (It) was one of the first important criticisms of the pessimistic view of English industrialisation, and it helped in the creation of a revisionary school that interpreted the industrial revolution more optimistically.'[13]

Hayek's Tribute to Ludwig von Mises

On 7 March 1956, at a party held in New York to commemorate the 50th anniversary of Mises's Doctorate, Hayek said: *'There has not been, and I don't expect that there ever will be, in my life another occasion when I have felt so honored and pleased to be allowed to stand up and to express on behalf of all those here assembled, and of hundreds of others, the profound admiration and gratitude we feel for a great scholar and a great man. It is an honor which I no doubt owe to the fact that among those available I am probably the oldest of his pupils.'*

This was the occasion when Hayek, speaking ten years after the death of Keynes, said *'I cannot resist the temptation to mention briefly one curious review which the book* [Mises's 1912 Theory of Money] *received. Among the reviewers was a slightly younger man by name of John Maynard Keynes, who could not suppress a somewhat envious expression of admiration for the erudition and philosophical breadth of the work, but who unfortunately, because, as he later explained, he could understand in German only what he knew already, did not learn anything from it. The world might have been saved much suffering if Lord Keynes's German had been a little better.'*[14]

Hayek, Ludwig von Mises and Fritz Machlup at the dinner held in von Mises's honour on 7 March 1956.

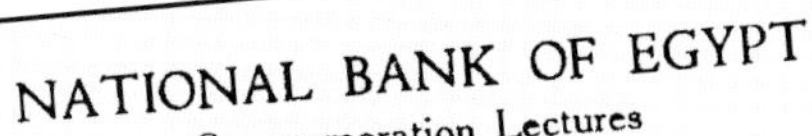

NATIONAL BANK OF EGYPT
Commemoration Lectures

The National Bank of Egypt take pleasure in announcing that they have arranged for a series of four lectures to be given at the Société d'Economie Politique, de Statistique et de Législation

by

Prof. F.A. VON HAYEK, F.B.A.
Dr. jur. et Dr. rer. pol. (Vienna)
D.Sc. Econ. (London)
Professor of Social & Moral Science
in the University of Chicago

The lectures will be delivered in English on 21st, 22nd, 23rd, and 24th February, 1955, at 6.30 p.m. and will deal with

"The Political Ideal of the Rule of Law"

You and your friends will be cordially welcome.

Hayek begins to write *The Constitution of Liberty*

Hayek was now embarking on his greatest book, *The Constitution of Liberty*, whose origins were *'an unforeseen but very pleasant fruit'* of his work on John Stuart Mill.

'In editing the correspondence with his wife, I had to omit most of the long letters Mill had written to her from the long journey to Italy and Greece he had taken for reasons of health in the winter and spring of 1854–55. It occurred to me that it might be interesting to repeat the journey after exactly a hundred years with the aim of producing a fully annotated edition of the letters. I succeeded in persuading the Guggenheim Foundation to give me a substantial grant to finance the journey....'

Hayek's series of four lectures in Egypt in 1955 were another starting point for his book *The Constiitution of Liberty*.

'Since [my wife and I] were able to travel by car so much more quickly than Mill had been able to travel once he got beyond the railways, and though we tried in general to be at the different places at the dates when Mill had visited them, we saved enough time to make from Naples a side trip to Egypt to deliver the lectures on The Political Ideal of the Rule of Law, *which I had been invited by the Bank of Egypt to give.'*[15]

The Political Ideal of the Rule of Law was a series of four lectures published in 1955 by the Bank on its 50th anniversary. Hayek says in the Preface that they should be regarded *'as an advance sketch of an argument which needs to be developed on a larger canvas'*.

Hayek later noted: *'These lectures, together with the constant preoccupation with Mill's thinking, brought it about that after our return to Chicago in the autumn of 1955, the plan for* The Constitution of Liberty *suddenly stood clearly before my mind.... I had before me a clear plan for a book on liberty arranged around the Cairo lectures. In the three succeeding years, I wrote drafts of each of the three parts of* The Constitution of Liberty, *revising the whole during the winter of 1958–59.'*[16]

'*At work on that book, about 1958*', wrote Hayek on the back of this photograph – an indication of the amount of time he had spent on *The Constitution of Liberty*, and in the Tyrol.

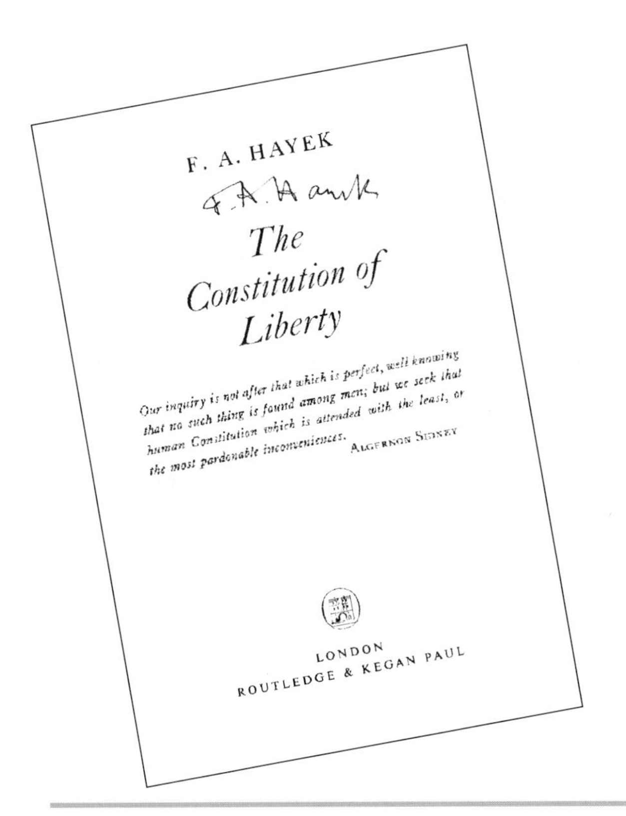

F. A. HAYEK

The Constitution of Liberty

Our inquiry is not after that which is perfect, well knowing that no such thing is found among men; but we seek that human Constitution which is attended with the least, or the most pardonable inconveniences.

ALGERNON SIDNEY

LONDON
ROUTLEDGE & KEGAN PAUL

The Austrian Tyrol always offered Hayek the solitude to write, so he did much of the work there. On his 60th birthday, 8 May 1959, Hayek was able to take the finished manuscript to his American publishers, the University of Chicago Press.

Published in February 1960, on the 100th anniversary of Mill's celebrated essay *On Liberty*, *The Constitution of Liberty* is a massive restatement of the principles and practice of liberalism in modern terms. It shows how society is a complex phenomenon, beyond the capability of any single mind to understand and therefore impossible to plan. Individual freedom is needed if it is to develop and be sustained, and any attempt to inhibit freedom will rob the social order of its unique ability to overcome new challenges and problems.

This review in the *Chicago Sunday Tribune* commended Hayek's 'far ranging work, encyclopedic in scope and disclosing a staggering learning, written with coolness and clarity'.

BOOK OF THE YEAR IN FIELD OF IDEAS

THE CONSTITUTION OF LIBERTY, by Friedrich A. Hayek [University of Chicago Press, 570 pages, $7.50].

Reviewed by George Morgenstern

This should be the book of the year in the field of ideas. The distinguished author of "The Road to Serfdom," which had a great impact 15 years ago, has pieced together the broken fragments of the tradition of liberty which developed in the western world but which has fallen into increasing disregard in the pursuit of alternative social orders.

Friedrich A. Hayek

Friedrich Hayek's far ranging work, encyclopedic in scope and disclosing a staggering learning, is written with coolness and clarity. It defines the nature and conditions of liberty, which is described as not merely one particular value but the source and condition of most mortal values.

• • •

Liberty is the condition in which coercion of some men by others, whether acting individually or as the state, is reduced as much as is possible in society. Liberty presupposes that the individual has some assured private sphere, in which he is his own master and can follow his own choice.

In free society, a monopoly of coercion is vested in the state, but it is tempered by the rule of law which places restraints upon state action and seeks to avoid discrimination by making all citizens subject to the same laws.

The state, however, has coercive powers in the form of taxation and compulsory service. It is in the provision of services and the application of policies supposed to enforce distributive or "social" justice that the state poses its principal threat to liberty, for when it does so it ceases to enforce general rules with equal effect on all citizens and necessarily engages in arbitrary discrimination between persons. That is, "A" is heavily taxed for the material improvement of "B."

• • •

Dr. Hayek finds that socialism is dead and the main danger now resides in the so-called "welfare state." The impulse of the dispensatory zealots leads them to overstep the bounds of the rule of law, which precludes all those measures which would be necessary to insure that individuals would be rewarded according to another's conception of merit, need, or desert rather than according to the value that their services have for their fellows.

So, tho socialism has been generally abandoned as a goal to be deliberately striven for, the author soberly feels that "it is by no means certain that we shall not still establish it, albeit unintentionally."

The book examines the legal framework required to support the liberal society, introducing Hayek's idea of the rule of law: treating people equally instead of as pieces in an economic chess game. It examines some of the economic institutions necessary to build a humane society with the minimum of coercion. As such, many readers with a background in practical affairs rather than in political philosophy have found it to be a useful introduction to Hayek's thought, and it is not therefore surprising that its influence has been so widespread.[17]

In London, *The Constitution of Liberty* was received with interest by a small group of enthusiasts gathered together in Antony Fisher's new Institute of Economic Affairs. The essays in the IEA's *Agenda for a Free Society* (1961) were derived from what the editor, Arthur Seldon, called Hayek's *magnum opus*, whose 'wide sweep takes in economics, political theory, philosophy, sociology, anthropology and jurisprudence'.

What ideals do *we* stand for?

THE CONSTITUTION OF LIBERTY

provides Americans with a modern restatement of the ideals of freedom that have guided, and must continue to guide, Western civilization if the free world is to win the battle for men's minds and hearts.

by Friedrich A. Hayek
author of
The Road to Serfdom
$7.50

Through your bookseller
UNIVERSITY OF CHICAGO PRESS
5750 Ellis Avenue, Chicago 37, Illinois
IN CANADA: *The University of Toronto Press, Toronto 5, Ontario*

MARCH 12, 1960 175

The IEA's contributors included (l to r): (back row) Arthur Seldon, Victor Morgan, Arthur Shenfield, Eric Nash, John Lincoln; (front row) Frederic Benham, F A Hayek, Graham Hutton.

F. A. HAYEK
COMMITTEE ON SOCIAL THOUGHT
UNIVERSITY OF CHICAGO
1126 EAST 59TH STREET
CHICAGO 37, ILL.

Chicago, December 1960

I would like my friends to know that the article entitled "New Nations and the Problem of Power" which appeared under my name in The Listener for November 3 reproduces the text of an unscripted recorded talk, almost in the form of an interview, which was extensively re-arranged by the editor and has not been seen by me before publication.

While I still hope that the general idea I wanted to convey comes through, I hope I will not be held responsible for the numerous obscurities, inaccuracies and even inconsistencies which the printed text contains.

F. A. Hayek

Celebrity has its problems: here Hayek complains to colleagues that his words have been twisted by a magazine published by the BBC in London.

9.40
NEW NATIONS
and the
PROBLEM OF POWER
by F. A. Hayek
Some thoughts on how emergent nations—and nations emerging from a period of dictatorship—might solve one of the key problems of democracy, that of limiting the powers of the State. F. A. HAYEK is the author of the recent book *The Constitution of Liberty*.
BBC recording

From the *Radio Times*, 5 November 1960.

Over the next 25 years, under the leadership of Ralph Harris as General Director and Arthur Seldon as Editorial Director, the Institute of Economic Affairs was to play a major role in promulgating Hayek's ideas in defence of the free society.

From the beginning of the Second World War, Hayek became increasingly deaf. Some years later he wrote: '*About the time of my move to Cambridge, and particularly in the new company of the High Table of King's College, I became rather depressed by my decreasing capacity closely to follow English conversations. It was only after the war, when I again visited German-speaking countries, that I discovered that it was not my understanding of English but my hearing that was deteriorating greatly. I had long been deaf on the left ear… but until above forty my hearing on the other ear was sharp enough to make up for it. It has since increasingly deprived me of the enjoyment of society and almost completely of the theatre, which at one time had been one of my regular pastimes…. it is in a large measure responsible for my appearing much more unsociable than I actually am.*'[18]

By the early 1960s, Hayek's years at Chicago were drawing to a close. He had suffered a severe, year-long depression soon after *The Constitution of Liberty* had been published. Now his family roots – and his love of the Austrian Alps – were pulling him back to Europe.

Hayek listening attentively at a meeting at Brown University in 1959, his deafness beginning to show.

So in 1962, Friedrich Hayek (now Professor Emeritus) left the University of Chicago. The tributes at his testimonial dinner on 24 May showed the high esteem in which he was held. His old friend and colleague Ludwig von Mises commented: 'We are not losing Professor Hayek entirely. He will henceforth teach at a German university, but we are certain that from time to time he will come back for lectures and conferences to this country. And we are certain that on these visits he will have much more to say about epistemology, about capital and capitalism, about money, banking and the trade cycle and, first of all, also about liberty. In this expectation we may take it as a good omen that the name of the city of his future sphere of activity is Freiburg. "*Frei*" – that means "free".'[19]

9 The prophet in the wilderness, 1962–1974

Hayek in Freiburg and Salzburg

After 31 years teaching in the English-speaking world, Hayek returned to Europe as Professor of Economic Policy at the Albert Ludwigs Universität at Freiburg im Breisgau, West Germany.

West Germany had achieved its 'economic miracle' since 1948 as a result of the famous 'bonfire of controls' ignited by its Finance Minister, Ludwig Erhard, who had been one of Hayek's guests at the second meeting of the Mont Pèlerin Society in 1948. His mentor was Walter Eucken, the liberal economist and head of the Freiburg School of economics.

Beyond West Germany, however, Hayek's ideas seemed increasingly out of fashion. In 1967, the British philosopher Anthony Quinton described him as 'the magnificent dinosaur' and Eric Hobsbawm said he was 'the prophet in the wilderness'. In the 1960s and 1970s Hayek was either ignored or publicly vilified. Becoming more and more depressed, his health began to suffer. It seemed as if he was in danger of becoming just another elderly academic, dismissed as a throwback to a bygone era.

When he retired in 1968, Hayek accepted an honorary professorship at the University of Salzburg, in his native Austria, where he lived until 1977 before finally settling back at Freiburg. Now, Hayek rarely travelled except to attend the meetings of the Mont Pèlerin Society. He only came to England to visit his family. He occupied an intellectual wilderness. But all this was to change dramatically in 1974.

Robbins and Hayek, reunited in 1961 at the wedding of Laurence and Esca Hayek.

Hayek at the University of Freiburg, 1962–1968

In 1962 Hayek left the University of Chicago, and returned to the German-speaking world following the unexpected offer of a Professorship at the Albert Ludwigs Universität at Freiburg im Breisgau in West Germany.

From 1962–1968 Hayek was *Professor der Volkwirtschaftslehre* (Professor of Political Economy) at Freiburg. He is seen here teaching in the 1960s.

'Much as I enjoyed the intellectual environment that the University of Chicago offered, I never came to feel as much at home in the United States as I had done in England. I also was much concerned about the inadequate provision for my and my wife's old age which that position offered me: a lump sum at a comparatively early retirement age (65). When I received in the winter of 1961–62 an unexpected offer of a professorship at the University of Freiburg im Breisgau, which not only was to run three years longer but also secured at least for me a moderate pension for life, I could have no hesitation in accepting the offer and have never regretted the move....

'I had, once again, to become an economist, but was able to concentrate in my teaching on the problems of economic policy, on which I felt I still had something of importance to say.

'We were very fortunate in finding an attractive apartment and particularly enjoyed the beautiful environment of the Black Forest. I was also fortunate to preserve almost to the end of that period at Freiburg my full energy and health and working capacity. And though after my seventieth birthday my powers began noticeably to decline ... they were on the whole very fruitful years.'[1]

Revisiting the campus many years later.

Freiburg University had been the intellectual home of Walter Eucken and his neo-liberal colleagues and was undoubtedly a congenial place to Hayek.

'Walter Eucken was probably the most serious thinker in the realm of social philosophy produced by Germany in the last hundred years.... [He] was a valuable friend for me. In the late 1930s, before the outbreak of the war, when I first acquired a car and made the trip from London to Austria by automobile, I regularly made a stopover in Freiburg just to visit Eucken and to keep in touch with him.'[2]

Wilhelm Röpke sent Eucken's *The Foundations of Economics* to Hayek in 1940: *'It made me realise for the first time what a towering figure Eucken was and to how great an extent Eucken and his circle embodied the great German liberal tradition, which had unfortunately become defunct... during the Nazi period.'*[3]

Walter Eucken (1891–1950), who had been at the first meeting of the Mont Pèlerin Society in 1947, was the leader of the 'Freiburg School' of economics, the pioneers of the *Soziale Marktwirtschaft*, the 'Social Market Economy', a phrase coined by another early MPS member, Alfred Müller-Armack.

The man most closely identified with the Freiburg School, Ludwig Erhard, joined the Mont Pèlerin Society at the second meeting. Erhard was the chief official responsible for the administration of economic affairs in the British and American military occupation zones in Germany after the war. In 1948, with Eucken and Röpke as his advisors, Erhard instituted his dramatic and famous 'bonfire of controls'. This led to the *Wirtschaftswunder'* ('economic miracle') over the next 40 years as West Germany's economy prospered in the post-war era. Erhard became the West German Minister of Economics 1949–1963, and Chancellor 1963–1966.

Ludwig Erhard (1897–1977) present at the second Mont Pèlerin Society conference in Seelisberg in 1949; his reforms initiated the German economic miracle.

'It must be admitted that Erhard could never have accomplished what he did under bureaucratic or democratic constraints. It was a lucky moment when the right person in the right spot was free to do what he thought right.... The freeing of prices was unbelievably successful. In the time that followed, there was in fact a more determined and conscious effort to maintain a free-market economy in Germany than in any other country.'[4]

At the 1956 Mont Pèlerin Society meeting in Berlin, on the general theme of 'The Challenge of Communism and the Response of Liberty', Erhard presented a key paper on 'The Economic Strategy of the Free World'.

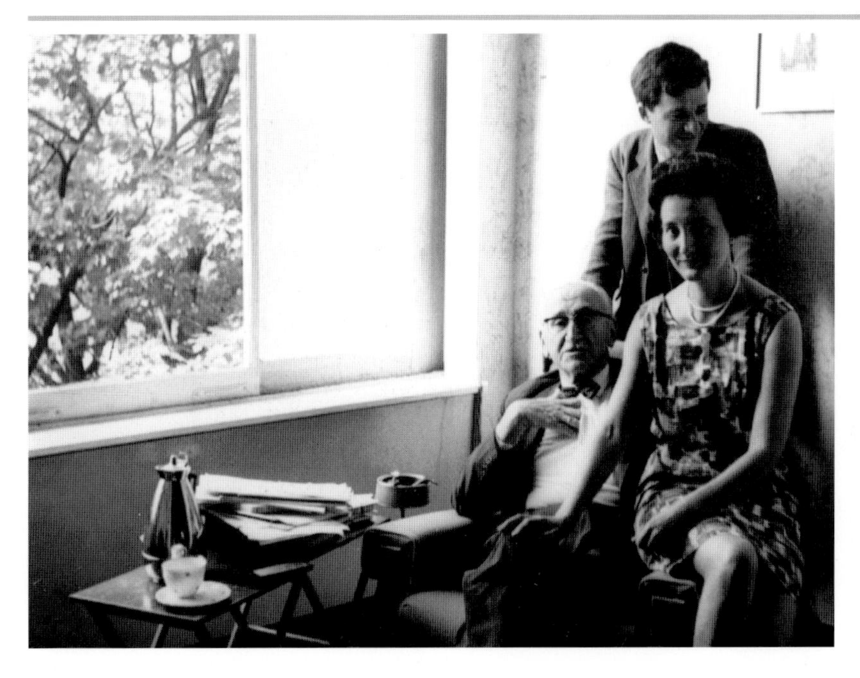

Laurence Hayek (son) and Esca Hayek (daughter-in-law) with Hayek in his apartment in Freiburg.

Hayek enjoying a visit to a medieval town in the Black Forest.

Despite the economic success of West Germany, in which he was now living and teaching, in the 1960s and 1970s Hayek's free-market ideas were largely out of fashion.

DER ERNST-REUTER-KREIS FREIBURG I. BR.
in Verbindung mit dem Colloquium Politicum der Universität Freiburg

gibt sich die Ehre, zum Vortrag mit Diskussion

PROFESSOR DR. F. A. VON HAYEK
in der Vortragsreihe
VORENTSCHEIDUNGEN DER WIRTSCHAFTSWISSENSCHAFT
– IHRE BEGRÜNDUNG UND IHRE PRAKTISCHE AUSWIRKUNG –

am Mittwoch, 16. Januar 1963, 20 Uhr c. t. im Hörsaal 2004, Kollegiengebäude II einzuladen.

CHRISTHARD SCHILLER

Invitation to a lecture given by Hayek in 1963 at Freiburg, with his notes.

Behind the Iron Curtain, dozens of communist governments imposed their will on millions of people. In many Western countries too, socialist governments followed the Keynesian nostrums and consensus. Writing in 1994, the radical historian Eric Hobsbawm recalled that, 'between the early 1940s and the 1970s the most prestigious and formerly influential champions of complete market freedom, eg Friedrich von Hayek, saw themselves and their like as prophets in the wilderness vainly warning a heedless Western capitalism that it was rushing along the "Road to Serfdom".'[5]

In the 1960s and early 1970s, partly because few chose to listen to Hayek's message, he travelled rarely. Although he had obtained British citizenship in 1938, he came to England only to visit his family: his daughter Christine in London and his son Laurence and family in Devon. But he continued hiking in his beloved Austrian Alps. He usually made his base at the Hotel Edelweiss at Obergurgl.

As he wrote on the back of this photograph, Hayek's climbing ended in 1967 but he continued to spend his summers in the mountains at Obergurgl.

With his family in Devon, England. From left to right: (on horses) grand-daughters Catherine Hayek and Ann Hayek; (middle row) Esca Hayek, Friedrich Hayek, Laurence Hayek; (front row) Tigger and Patch the dogs, grandson Crispin Hayek.

Most of Hayek's travelling was to attend Mont Pèlerin Society meetings and debate with friends of like minds. In the 1960s and 1970s, general meetings of the society were held in Kassel (1960), Turin (1961), Knokke (1962), Semmering (1964), Stresa (1965), Vichy (1967), Aviemore (1968), Munich (1970) and Montreux (1972).

Hayek was still revered among this circle of friends and other sympathetic colleagues in academe. In public life, however, he was virtually ignored or often actually publicly vilified. The Oxford philosopher Anthony (later Lord) Quinton referred to him as the 'magnificent dinosaur'.

Hayek with his daughter Christine.

This time in the wilderness had its effect; often he seemed close to despair. Depressed, his health suffered. He was in danger of becoming just another elderly academic, dismissed as a throwback to a bygone era. As he later wrote: '*It will not be easy for future historians to account for the fact that, for a generation after the untimely death of Maynard Keynes, opinion was so completely under the sway of what was regarded as Keynesianism....*'[6]

The Institute of Economic Affairs

But Hayek's ideas *were* percolating, helped by a tiny new institute in London.

Hayek with Antony Fisher, pictured outside the Reform Club in London.

Antony Fisher, a former Battle of Britain pilot, had first met Hayek in 1946 at the London School of Economics. As Fisher wrote, 'Hayek first warned against wasting time – as I was then tempted – by taking up a political career. He explained his view that the decisive influence in the great battle of ideas and policy was wielded by the intellectuals… his counsel was that I should join with others in forming a scholarly research organisation to supply intellectuals in universities, schools, journalism and broadcasting with authoritative studies of the economic theory of markets and its application to practical affairs.'[7]

By 1955, Fisher was running a successful farm business and was able to provide the necessary funds to put Hayek's advice into practice. He established the Institute of Economic Affairs and invited a young economist, Ralph Harris, to run it on £50 a month from a small office in London.

Under the leadership of Ralph (later Lord) Harris and editorship of Arthur Seldon, the IEA produced many publications, including a series of Hobart Papers – named after the street in which it was located.

The Institute directed a barrage of ideas into academe and the political, literary and journalistic establishments. Hayek himself added to this intellectual artillery over the years, writing numerous pamphlets for the Institute including *Denationalisation of Money* in which he proposed competing private currencies and abandoned his long-held belief that control of the monetary system was a proper function of government.

At last this mild-mannered and rather reserved academic had found what was to become a very effective channel through which to influence others.

Arthur Seldon (left) and Ralph Harris at an early meeting of the Mont Pèlerin Society.

Sir Karl Popper (left) with Hayek.

F. A. Hayek

STUDIES IN
PHILOSOPHY, POLITICS
AND
ECONOMICS

A CLARION BOOK
PUBLISHED BY SIMON AND SCHUSTER

Hayek published *Studies in Philosophy, Politics, and Economics* in 1967. Dedicated by Hayek to his *'old friend'* Karl Popper (to whom he said he owed many *'intellectual debts'*), the book contains 25 essays, articles and speeches from a 20-year period. Hayek confirmed his multidisciplinary approach *'to the public issues of our time'*, writing in the Preface:

'This volume contains a selection from the work of the last twenty years or so of an economist who discovered that if he was to draw from his technical knowledge conclusions relevant to the public issues of our time, he had to make up his mind on many questions to which economics did not supply an answer.'

The table of contents gives an indication of the remarkable breadth of Hayek's intellectual interests, with essays on: kinds of rationalism; the theory of complex phenomena; the results of human action but not of human design; the legal and political philosophy of David Hume; the moral element in free enterprise; full employment, planning and inflation; unions, inflation and profits; the corporation in a democratic society; and a re-assessment of *The Road to Serfdom* after 12 years.

Hayek skiing at Aspen, Colorado in 1969 – at the age of 70!

On his 70th birthday in 1969, the Faculty and students at Freiburg honoured Hayek (then Professor Emeritus) with gifts and a candlelight parade.

Though governments might be ignoring his conclusions, at least some in the academic world recognized his lifetime achievements as a social theorist. When Hayek retired in 1968, he accepted an honorary professorship at the University of Salzburg in his home country of Austria.

In his study at Salzburg, 1974.

In retirement, he was awarded many honours for his lifetime's work in philosophy, political science and economics. His reputation had already brought him an honorary doctorate from Rikkyo University in Tokyo, Japan in 1964. In 1972 the London School of Economics made Hayek an Honorary Fellow, while in 1977 the Hoover Institution at Stanford University appointed him Visiting Fellow.

1971: Hayek receiving the distinction of Honorary Senator at the University of Vienna.

1974: Receiving an honorary doctorate from the University of Salzburg.

Hayek on '*spontaneous order*'

Shortly after he left Freiburg, Hayek went through a severe depression which caused *'inner trembling… which intellectually disabled me'*. It contributed to what he called *'my miserable state in the early 1970s'*.[8] So before long he was back in Germany and embarking on another big project.

'Soon after settling down at Freiburg, I started work on what grew into the project of a rather ambitious book on law, legislation and liberty, intended as a kind of supplement to The Constitution of Liberty.*'*[9]

His poor health during this period still caused Hayek and his friends much concern. He was able to publish the first volume of his new book, *Law, Legislation and Liberty* in 1973 but for some time he said that he hardly dared hope that he would live to complete the trilogy.[10]

Hayek wrote in the Introduction that:

'What led me to write another book on the same general theme as The Constitution of Liberty *was the recognition that the preservation of a society of free men depends on three fundamental insights which have never been adequately expounded and to which the three main parts of this book are devoted.'*

The first was that there could be order, without commands – a *'self-generating or spontaneous order'*. The rules which allow that to work are quite different to the tenets of a deliberate organisation.

Hayek bought many of his books from the Alternative Bookshop in London's Covent Garden. He is seen here with John Wood (right), and Chris Tame, Manager.

Hayek signing copies of *Law, Legislation and Liberty* at the Alternative Bookshop.

The second was that 'social justice' has meaning only within an organisation, and is *'wholly incompatible'* with an open society.

The third was that the prevailing idea of democracy, in which the same representative body lays down the rules of just conduct and directs government, eventually perverts the spontaneous order of a free society into *'a totalitarian system'*.

The *Economist* commented that Hayek 'in this, as in his earlier writings, presents a powerful challenge to current trends in social theory and policy.'

Yet as 1974 began, the retired, elderly professor had no notion as to what was in store for him very shortly on the international stage.

10 The power of ideas 1974–1992

From the Nobel Prize to *The Fatal Conceit*

In 1974, unexpectedly, the Nobel Prize in Economics was awarded to Friedrich Hayek. The award helped rejuvenate the ailing scholar and he enjoyed a renewed burst of energy. Now well into his seventies, at last he completed his trilogy *Law, Legislation and Liberty*.

Following the succesful example of the Institute of Economic Affairs in London, dozens of new liberal economic policy institutes were formed and flourished in countries from Austria to Australia, Peru to Canada, France to the United States – drawing their inspiration from the works of Hayek, Friedman and other members of the Mont Pèlerin Society. This, too, was becoming a truly international gathering of academics, many of whom now had impressive reputations in their own rights, with seven of its members being awarded the Nobel Prize in Economic Science.

By the 1970s, Keynesianism was coming under suspicion as postwar inflation soared while unemployment refused to fall. Hayek's approach provided a solution to this puzzle, his proposals urging cuts in state spending, deregulation, and curbing the monopoly power of trade unions. As a result, he was widely acknowledged as one of the major intellectual roots of both the Thatcher and Reagan administrations. After years of rejection, he was at last being taken seriously.

Hayek's last book, published in 1988 and entitled *The Fatal Conceit: The Intellectual Error of Socialism*, challenged the very foundations of collectivism. A year later, the Berlin Wall would be torn down by the peoples whom it had imprisoned for so long.

Having lived just long enough to see the rediscovery of freedom in much of Eastern Europe, Professor Friedrich August Hayek died in Freiburg im Breisgau, in the newly re-unified Germany, on 23 March 1992, in his 93rd year.

Described by *The Economist* as 'the century's greatest champion of economic liberalism', he left a huge intellectual legacy that will endure for all time.

Hayek awarded the Nobel Prize in Economics, 1974

Hayek receiving his award from the King of Sweden.

When in 1974 Hayek was awarded the Nobel Prize for Economics, jointly with his old intellectual adversary Gunnar Myrdal, the *enfant terrible* of the economics profession had at last become one of its grand old men.[1]

The Nobel Prize was a remarkable achievement for one who had always assumed that his warnings against the intellectual drift to socialism would be so unpopular as to bar him from such honours, particularly from the Swedish Academy of Sciences.[2]

'The Nobel Prize was a complete surprise to me. I didn't approve of Nobel Prizes for economists – until they gave it to me, *of course! Of course there is a very big advantage to fame: people suddenly listen to you.'*[3]

Introducing the two prize winners in Stockholm on 11 December 1974, Professor Erik Lundberg of the Royal Academy of Sciences said that: 'Hayek's contributions in the fields of economic theory are both deep-probing and original. His scholarly books and articles during the 1920s and 1930s sparked off an extremely lively debate. It was in particular his theory of business cycles and his conception of the effects of monetary and credit policy which aroused attention.... Perhaps in part because of this deepening of business-cycle analysis, Hayek was one of the few economists who were able to foresee the risk of a major economic crisis in the 1920s, his warnings in fact being uttered well before the great collapse occurred in the autumn of 1929.'

Turning to Hayek's more recent work, Professor Lundberg said: 'It is above all the analysis of the viability of different economic systems which is among Professor Hayek's most important contributions to social science research. From the middle of the 1930s onwards, he devoted increasing attention to the problems of socialist central planning.... His guiding criterion in assessing the viability of different systems refers to the efficiency with which these systems utilise the knowledge and information spread among the great mass of individuals and enterprises. His conclusion is that it is only through a far-reaching decentralisation in a market system with competition and free price formation that it is possible to achieve an efficient use of all this knowledge and information.'[4]

Friedrich Hayek standing in front of the portrait of Alfred Nobel with (l to r) his son Laurence, daughter-in-law Esca and daughter Christine.

Hayek, with his family, celebrating his award. Mrs Hayek is seated opposite her husband.

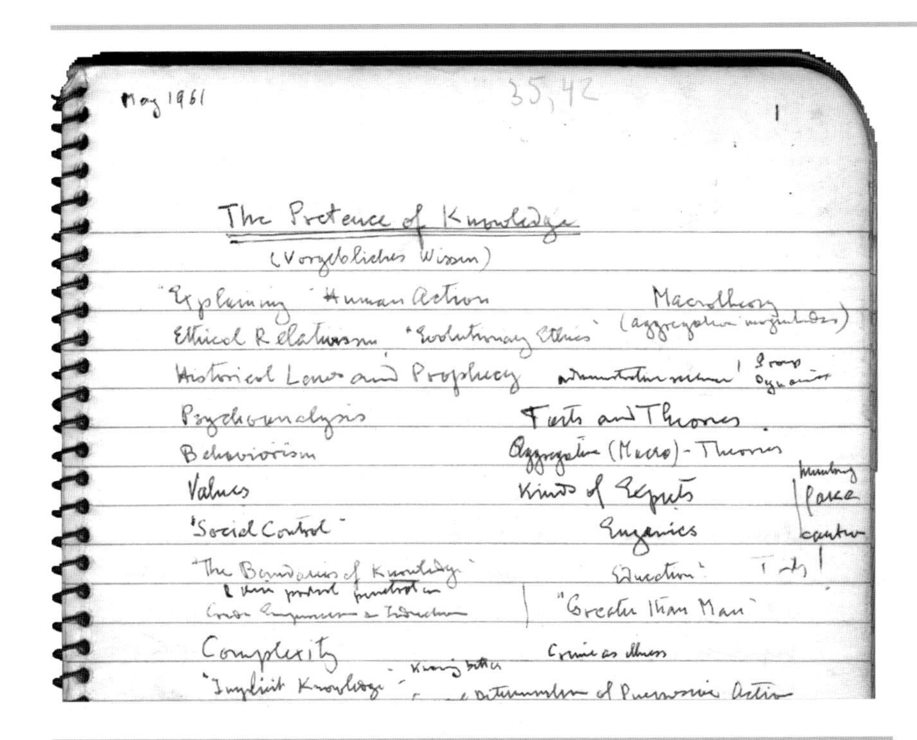

Hayek had worked on the '*pretence of knowledge*' idea many years before at the University of Chicago. This 1961 draft in his notebook has never previously been published.

In response, Hayek's Nobel Memorial Lecture attacked economists for what he called *The Pretence of Knowledge.* He chided his profession with the words: '*On the one hand the still recent establishment of the Nobel Memorial Prize in Economic Science marks a significant step in the process by which, in the opinion of the general public, economics has been conceded some of the dignity and prestige of the physical sciences. On the other hand, the economists are at this moment called upon to say how to extricate the free world from the serious threat of accelerating inflation which, it must be admitted, has been brought about by policies which the majority of economists recommended and even urged governments to pursue. We have indeed at the moment little cause for pride: as a profession we have made a mess of things.*'

He then explained exhaustively how he felt that the greatest mistakes of recent economic policy stemmed from the misapplication of supposedly 'scientific' approaches where these were in fact inappropriate.

He concluded: '*The recognition of the insuperable limits to his knowledge ought indeed to teach the student of society a lesson of humility which should guard him against becoming an accomplice in man's fatal striving to control society – a striving which makes him not only a tyrant over his fellows, but which may well make him the destroyer of a civilisation which no brain has designed but which has grown from the free efforts of millions of individuals.*'[5]

The award of the Nobel Prize returned Hayek to the headlines and he received extensive press coverage internationally. Cuttings collected in his own album reflect both favourable and unfavourable views.

Daily Telegraph 14.ix.78. — Hayek

Mrs Thatcher and 'mad professor' attacked by Foot

By NICHOLAS COMFORT, Political Staff

MR FOOT, leader of the Commons, last night accused Mrs Thatcher of being in clutches of a "mad professor" with a dangerous philosophy involving a return to the mass unemployment of the 1930s.

He was ... Conservative ... stituency of ... Mr Foot sa... referring to Sir ... "though obvio... under that hea... Prof. Friedrich ... Prof. Hayek, 7... tired and lives on ... has not been ret... Thatcher as an ad... writings, during ...

Wirtschafts-Nobelpreis für Hayek und Myrdal

C. G. STOCKHOLM, 9. Oktober. Der Nobelpreis für Wirtschaftswissenschaft 1974 ist dem Österreicher Friedrich von Hayek und dem Schweden Gunnar Myrdal verliehen worden. Die beiden Nationalökonomen, so begründet die schwedische Akademie der Wissenschaften ihre Entscheidung, werden für ihre „bahnbrechenden Arbeiten auf dem ... der Geld- und Konjunktur- ... ihre tiefgründigen Ana- ... Abhängigkeit ...

PAUL JOHNSON — THE TUESDAY COLUMN

Who's afraid of the formidable Fritz?

Razor sharp ... Fritz Hayek.

... have all the Left ... tectuals gone? Bud- ... ey have vanished— ... ago, silent or lying ... hangs just mentally ... And, of course, ... hanged ... this is ...

On America's East Coast, the young have turned away from the once-trendy New York Review of Books and read Commentary, another citadel of anti-collectivism.

... has a mind like a razor. His Road to Serfdom is read all over the world, and his monumental study of modern politics, Law, Legislation and ...

Democratic Ideal Prevail, ... new volume of Right-... essays, published on Thurs... called Confrontation: Will The Open Society Survive to 19... contributors include ... leading academics as the soci... gist Professor Julius Gould ... the historian Professor Nor... Gash. But the book also featur... remarkable newcomer to ... field of polemics: the Duke ... Edinburgh. In an essay, Intelle...

10.X.74 THE NEW YORK TIMES

Unretiring Economist Whose Energy Defies Age

Friedrich August von Hayek

Who the heck is Hayek?

By VICTOR KNIGHT, Political Editor

AN Austrian-born professor is the shadowy figure behind a flaming row inside the Tory Party.

He is seventy-seven-year-old Friedrich von Hayek, a Nobel prize-winning economist whose free-for-all doctrines may become official Conservative policy.

WE ARE ALL HAYEKIANS NOW

IT is not very easy to find a thorough-going Keynesian these days except perhaps in the backrooms of the C.B.I. and the T.U.C.

The academics have got their heads well down, the Treasury no longer believes in anything and when after the April Budget ...

... are being visibly fulfilled. Third, he has throughout his life been an arch anti-inflationist.

At a time when the Government's "Attack on Inflation" White Paper ends with these chilling words: "This is a plan to save our country. If we do not achieve ...

... a choice between only three possibilities: To allow a rapidly accelerating ...

PATRICK HUTBER

Nobel Prize winners disagree on economics

By Tim Congdon

The 1974 Nobel Economics was join... in Stockholm ...

CONTRASTED NOBELMEN

BLACK AND WHITE, God and the devil, chalk and cheese —these are conventional opposites. This year's Nobel ... economics has been split between Professors ... and FRIEDRICH VON HAYEK ... These distinguished men ... the economic field for ... chosen principally to operate; with the ability to communicate ... people: a grave and at times ... future of man, considered not ... but as a whole and complex ... two savants more sharply ... found. It is hard to admire ... The Royal Swedish Academy ... somehow—or is there no corporate ...

... are not adjectives to be used ... Which you prefer— ... on your world outlook ... Do you regard poverty ... or international, as the ... and the greatest danger to ... your man. Or do you ... law as the greatest of earthly ... however desirable in itself, ... A HAYEKIAN. You cannot easily ... and, whichever you choose.

Arthur Shenfield looks at the life of Professor Hayek, a Nobel prizewinner, whose ideas form an important strand of the present economic debate in Britain

A prophet coming into his own?

To be a highly respected figure in the scholarly world for half a century, to teach and write prolifically in all that ... on subjects close to the ... of the general ...

... ial political and legal theory. Thus, what is perhaps his greatest work, The Constitution of Liberty, is in the latter field (though it displays the hand of a political philosopher who is also an economist).

It is a work which will rank imperishably among the classics of political philosophy ... perception of ... line of ...

... political philosophy and legal theory (The Road to Serfdom, 1944; Individualism and Economic Order, 1948; The Counter-revolution of Science, 1952; The Constitution of Liberty, 1960; Studies in Philosophy, Politics and Economics, 1965; Law, Legislation and Liberty, 1974, etc).

The essence of Hayek's teaching in economics, which is at last attracting the attention of the general public, is to be found in the earliest works listed above.

It was not then worked ... with comple...

The Nobel Prize for economics

AN APPRECIATION BY LORD ROBBINS

Financial Times 11.X.74

Professor Hayek denounces Marx and Freud

Times 15.ix.78

By David Nicholson-Lord

The twentieth century will ultimately be regarded as an age of superstition, associated chiefly with the names of Karl Marx and Sigmund Freud, Professor F. A. Hayek, the economic philosopher and Nobel prizewinner, predicted last night.

Delivering ...

In 1949 Hayek had written in *The Intellectuals and Socialism* that '*If we can regain that belief in the power of ideas which was the mark of liberalism at its best, the battle is not lost*'.[6] Twenty-five years later, after the Nobel Prize, his views and opinions were in demand all over the world.

Through his constant barrage of 'Letters to the Editor' over half a century, Hayek kept his name in front of a significant public and in front of what he called, '*part maliciously and part facetiously, the "second-hand dealers of ideas"*', the journalists and leader writers who form '*a group of decisive importance because it determines what the masses think*'.[7]

~~THE TIMES 21 XII 76~~

The German economy

From Professor F. A. Hayek

Sir, You report in your issue of December 18 that Mr Roy Jenkins is concerned about the solution of the fundamental problems of the unbalance caused by Germany being in a position quite different from that of France, Britain and Italy. The solution is the simplest, most obvious possible. That the latter three countries abandon their socialist policies, and they will in a few years be where Germany has arrived now by this method.

Yours sincerely,
F. A. HAYEK,
Salzburg,
Austria.

Benefits of industrial destruction

D.T. Aug 8 '75

From Prof. F. A. HAYEK

SIR—I find it difficult to trust "the first comprehensive, on-the-record Press interview . . . since his February, 1974, election" which *Time* magazine claims to have been granted by the Prime Minister.

Is it really conceivable that a British Prime Minister is such a fool as seriously to maintain that "Germany, of course, had the advantage of the total destruction of their industrial equipment, while we were struggling with old looms, old machine tools and the rest"? If that is so, why not proceed at once to the total destruction of most of Great Britain's industrial equipment?

F. A. HAYEK
Salzburg.

... that a party that keeps a socialist government in power clearly has lost all title to the name "Liberal". Certainly no liberal can in future vote "Liberal".

Yours faithfully
F. A. HAYEK
D-78 Freiburg i Brg.

LT EDITOR DAILY TELEGRAPH
135 FLEET STREET
LONDON EC
TELEX 22874–6

ONE MUST STILL DOUBT WHETHER ANY MAN CAN DO MUCH GOOD AT HEAD OF PARTY CONSISTING OF TRADE UNION MONOPOLISTS AND DOCTRINAIRE INTELLECTUALS STOP BUT ONE OUGHT TO GIVE UNSTINTED PRAISE TO AN HONEST MAN WHO FACES THE FACTS AND TELLS HIS PARTY THE UNPOPULAR TRUTH ABOUT THE CAUSES OF UNEMPLOYMENT AND THE NECESSITY OF PROFITS STOP HOWEVER MUCH ONE MAY DISAGREE ON MANY POINTS MR CALLAGHAN SURELY DESERVES THE PRIZE FOR BEING THE MOST COURAGEOUS BRITISH PRIME MINISTER IN A LONG TIME F A HAYEK

DECLARATION BY SENDER OF FOREIGN LETTER TELEGRAM OR PRESS TELEGRAM
I hereby certify that the particulars of the above telegram contain only English language and that the text does not bear any meaning other than that which appears on the face of it.
Sender's Signature

F. A. HAYEK
ORIENTAL HOTEL
BANGKOK

Date 30 . 9 . 76

Economist June 4, 1932, p. 1243

FOREIGN EXCHANGE RESTRICTIONS.

TO THE EDITOR OF THE ECONOMIST.

SIR,—The admirable article on Foreign Exchange Restriction in Danubia in your current issue tempts me to seek the hospitality of your columns for a few remarks upon a point which has not, as yet, attracted much attention in this country.

Anybody familiar with the present plight of Central Europe can have little doubt that the exchange restrictions have not only failed to achieve the desired object, but that they have actually done more than any other single factor to aggravate the situation. And it is certain that the dwindling of trade in this area—which is the main concern of the rest of the world—is chiefly due to this one factor. Informed opinion in Germany and Austria is unanimous in condemning these restrictions, in even stronger terms than were employed in your article. This fact would probably have prevented their imposition, or at least have led to their early removal, had they been instituted by over-eager bureaucrats, if it had not been for the curious factor to which I desire to draw attention.

If one approaches any responsible person in the countries concerned and protests against these restrictions, one is almost certain to be informed that the establishment of this régime was due not to their own choice but was done on the advice—or perhaps one should say request—of the representatives of the foreign creditors. In Austria—with whose position I am more familiar—it would have been considered indiscreet to mention this fact in public until recently; but, in a recent speech, Chancellor Brüning has mentioned the fact that in Germany, too, exchange restriction was forced upon an unwilling government by the representatives of the foreign creditors.

Now, in discussions of this kind, the creditor nations were, so far as one knows, represented by supposed experts. But is it credible that, at the present time, experts should give advice of so ruinous a nature? This would imply that we have not only failed to learn anything during the past ten years, but that we have also forgotten what we had already learned. In the spring of 1922 the Finance Commission of the International Economic Conference adopted a resolution that—

"All artificial control of operations in exchange, whether by requiring a licence for transactions in exchange, or by limiting the rates at which transactions may be effected, or by discriminating between the different purposes for which the exchange may be required, or by preventing free dealings in forward exchange, is futile and mischievous, and should be abolished at the earliest possible date."

Is it not time to ask who are the "experts" who, by advice which is in flagrant conflict with all sound scientific teaching and with the published opinion of a body of experts who had to accept responsibility for their advice, have inflicted this inestimable harm not only upon Central Europe but also upon the world as a whole?—Yours, &c.,

London School of Economics,
Houghton Street, W.C.2.
May 31, 1932.
F. A. VON HAYEK.

A selection of Hayek's many 'Letters to the Editor'.

Looking back on his period of ill health, Hayek told the Carl Menger Society in London: *'Some years ago I tried old age, but discovered I didn't like it'*. In 1973 he had been proposed, unsuccessfully, for the honorary position of Chancellor of the University of St Andrews in Scotland, but the electorate had decided that he was too old and frail for the job. Hayek laughed uproariously nine years later when, during a visit to the Adam Smith Institute, he was told that he was the only candidate still alive.

Professor F.A.Hayek
D-78 Freiburg i.Brg.
Urachstrasse 27
Tel. 060761/77216

PROVISIONAL PLAN OF MOVEMENTS APRIL TO NOVEMBER 1978

At Freiburg until June 25 except
May 12-23 in ENgland of which 15-19 London (Reform Club, Tel930 9374)
(Hobhose Lecture at London School of Economics 17t
and May 28-31 at Bonn to receive Pour le merite
At Lindau (Lake Constance) June 26-30 (meeting of Nobel Laureates)
At Obergurgl July 1 to August 20 (A-6456 Obergurgl, Austria, Tel.
Hotel Edelweiss, Tel.0043/05256/223-4)
At Freiburg August 21-30
Dept Frankfurt for New Delhi August 30
At New Delhi August 31 - Sept. 2, Intercontinental Hotel
Dep. New Delhi for Hongkong Sept.2
At Hongkong (Mont Pelerin Society) Sept.3-10
In South Korea Sept. 10-17
In Japan (chiefly Kyoto) September 17 to October 15
At Stanford. Cal. cca October 16 to November 15
In England probably Sept.15-26

After the Nobel Prize, Hayek was fêted around the world and his travel plans show an extraordinary burst of activity. This itinerary shows Hayek's remarkable energy as he approached his 80th birthday.

Eamonn Butler, in his intellectual biography of Hayek, wrote: 'The Nobel Prize was an achievement which gave Hayek a renewed burst of energy and health, and he began writing and lecturing even more widely than before. It was with a mixture of relief and delight that applause drowned out the end of a telegram from the absent Hayek which was read to the Mont Pèlerin Society meeting in Hillsdale, Michigan, in 1975. All that was audible was "*I have just completed Volume 2 of...*".'[8]

That second volume of *Law, Legislation and Liberty* was published in 1976, and within a few years Volume 3 had also appeared. The journal *Philosophical Studies* commented: 'This three-volume series promises to be the crowning work of a scholar who has devoted a lifetime to thinking deeply about society and its values'.

Hayek's candid admission in the Preface to Volume 3 gives hope to all authors. *'Again unforeseen circumstances have delayed somewhat longer than I had expected the publication of this last volume of a work on which I had started more than seventeen years ago.... But while I believe I have now more or less carried out the original intention, over the long period which has elapsed my ideas have developed further and I was reluctant to send out what inevitably must be my last systematic work without at least indicating in what direction my ideas have been moving'*.

Law, Legislation and Liberty developed Hayek's earlier work exploring the legal arrangements which are necessary in the free society. It shows how the roots of social life can be found in human evolution (rather than in conscious planning), exposes the lack of precision of 'social' or redistributive justice, and puts forward suggestions for a constitutional arrangement which would keep down the arbitrary powers of government authority.

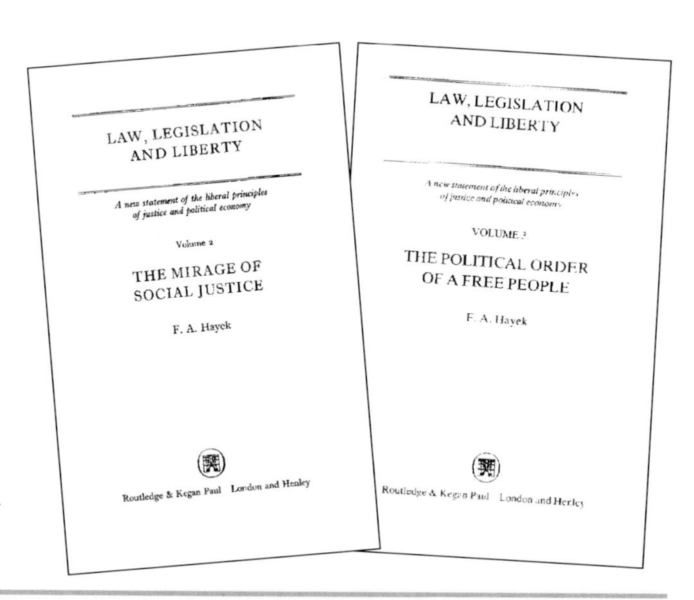

Hayek's legacy: the Mont Pèlerin Society

Hayek left an enduring legacy in the Mont Pèlerin Society, which grew to become a truly international gathering of academics, many of whom had massive international reputations in their own right. General meetings were held in Brussels (1974), St Andrews (1976), Hong Kong (1978), Stanford (1980), Berlin (1982), Cambridge (1984), St Vincent (1986), Tokyo (1988), Munich (1990), Vancouver (1992), Cannes (1994), Vienna (1996) and Washington, DC (1998).

Presidents of the Society include many distinguished champions of liberalism:

F. A. Hayek, 1948–1960
W. Röpke, 1960–1961
J. Jewkes, 1962–1964
F. Lutz, 1964–1967
D. Villey, 1967–1968
F. Lutz, 1968–1970
M. Friedman, 1970–1972

(l to r) Pascal Salin, Edwin J. Feulner, John Gray and Friedrich Hayek at the 1984 Cambridge meeting.

(l to r) William Hutt, Friedrich Hayek, Arthur Shenfield.

On the back of this photograph, Hayek has written: "*This was my formal 80th birthday portrait*'.

A. A. Shenfield, 1972–1974
G. G. Leduc, 1974–1976
G. J. Stigler, 1976–1978
M. Ayau, 1978–1980
C. Nishiyama, 1980–1982
R. Harris, 1982–1984
J. M. Buchanan, 1984–1986
H. Giersch, 1986–1988
A. Martino, 1988–1990
G. Becker, 1990–1992
R. M. Hartwell, 1992–1994
P. Salin, 1994–1996
E. J. Feulner Jr, 1996–1998

At the St Andrews meeting of the Mont Pèlerin Society, 1976. *Right*: Hayek on the pier wall at St Andrews delivering a eulogy to Adam Smith on the 200th anniversary of the publication of *The Wealth of Nations*.

Below: Front row (l to r): Milton & Rose Friedman, Gaston Leduc, Friedrich Hayek; Back row (l to r): Martin Anderson, (Adam Smith), Ralph Harris.

Max Hartwell, historian of the Society, with Friedrich Hayek.

Hayek addressing one of the last General Meetings of the Society that he attended, Cambridge, 1984.

Nobel Laureates from the Mont Pèlerin Society

Out of its comparatively small membership of never more than 500, no less than seven of its members have been awarded the Nobel Prize in Economic Science. This total includes four of the economists who were at the original meeting in 1947, including Friedrich Hayek himself in 1974, followed by:
Milton Friedman (USA), 1976
George Stigler (USA), 1982
James Buchanan (USA), 1986
Maurice Allais (France), 1988
Ronald Coase (USA), 1991
Gary Becker (USA), 1992.

(l to r) James Buchanan, Milton Friedman, George Stigler.

Maurice Allais.

Gary Becker.

Ronald Coase.

Hayek and the institutes worldwide

Hayek's ideas also spread through the growing network of free-market institutes, because he recognised the crucial role that Antony Fisher's Institute of Economic Affairs played in changing opinion in Britain. In a 1983 interview, he said:

'It has taken a long time to prove its success.... I now think it has become the most powerful maker of opinion in England. By now, book shops usually have a special rack of Institute of Economic Affairs pamphlets. Even people on the left feel compelled to keep informed of the Institute's publications.'[9]

New institutes, such as the Adam Smith Institute in London (of which Hayek was Chairman), the Fraser Institute in Vancouver and the Heritage Foundation in Washington DC, all promote liberal economic ideas in their own way, achieving a powerful impact on the media and the public in their respective countries around the globe.

1985: Ralph Harris and Arthur Seldon bid Hayek farewell outside the IEA in London.

The IEA's former Marketing Manager, John Raybould, with Nobel Laureate Ronald Coase at an IEA display at the LSE in 1981.

Friedrich Hayek and Ralph Harris at the IEA.

In the first of two articles, JOHN CUNNINGHAM contemplates the growing mini-industry of political policy units

The think-tanks' role in the battle for Britain

"... the influential Adam Smith Intitute ... says that private enterprise should take over town hall services." *The Daily Express, London.*

Pictured below: Dr. Eamonn Butler (left) and Dr. Madsen Pirie.

vate supply of ancillary health services. The

tion of the rise of "Quango" government, an

On the Intellectual Ramparts

New think tanks are advocating as well as incubating

As the day drew near for a H
aid to the Nicaraguan
Heritage Foundation massed
behalf of the rebel troops. In
roon auditorium just a few bl
Capitol, it held an all-day se
gressional staffers. The gu
two top *contra* officers and
opposition journalist. A w
tage issued a brisk nine-pa
Nicaragua's Terrorist Co
of which were distributed
Congressmen and to targ
bers. Heritage's pro-*con*
The reign of the pens
pipe-smoking Washingto
is under assault. These
search institutions, which
the first decades of this
being upstaged by group
al crusaders that helped
publicans the party of id
the way for Ronald Re
The new "advocacy t

Dry books surprise hits for institute

BY JOHN CLARKE

VANCOUVER

If having at least three bestsellers after only five years in the business is the mark of success, then Vancouver has one of the most successful new publishing houses in the country. And, surprisingly, the books are all about dry-as-bones economics.

Since it opened shop in 1974, the Fraser Institute has produced three books with more than 10,000 in sales, which is bestseller country in Canada. One was about wage and price

BRITISH COLUMBIA

The 'age of Hayek'

By the late 1970s and early 1980s, many strands in Hayek's life were coming together. In 1983, reviewing his life-long crusade in defence of the free society and the market system, he reflected: *'An evil fate befell German efforts to defend the ideal of liberty in general and in the field of economics in particular, with the result that today I am almost the only survivor of a generation that set out in the wake of the First World War to devote all its energies to the preservation of a civilised society.... Primarily under the impetus of Ludwig von Mises, a very small group of us dedicated ourselves to this task in the post war years.... [Today] I am much more confident... about the prospects of a worldwide restoration of a market economy.'*[10]

Hayek's optimism about the future was well founded. Against a background of high inflation and high taxation in the industrialised nations, his work provided a scholarly foundation for a dramatic change in the direction of politics – on both sides of the Atlantic.

Now, new politicians were coming to power and promising to roll back the frontiers of the state as an alternative to the old policies of interventionism and Keynesian demand management.

WHERE KEYNES WAS WRONG

The republication of an anthology of Professor F. A. Hayek's writings on inflation and Keynes is an opportunity to reassess the impact of a man whose greatest influence has come towards the end of his life. The third quarter of this century has been described as "the age of Keynes" not just because of the influence which he had on economics and public policy but also because it was above all a time when his predictions of the beneficial effects which would flow from his policies seemed to be triumphantly justified.

In terms of the economic problems now facing us, the current period might more accurately be termed "the age of Hayek". Inflation has, on occasion, spiralled out of control and is still too high. Attempts to remove unemployment by stimulating demand have become more and more obviously ineffective. Professor Hayek tends to be linked with the other school of opponents of Keynes who are generally described as monetarists and whose chief exponent is Professor Friedman. There are in fact important differences in the economic analysis which underlies his position from the views of traditional monetarism, even though many of the political and social judgments are the same.

At the heart of Professor Hayek's disagreement with Keynes is a concern for the relative movements of prices in the economy rather than the performance of overall aggregates. In practical terms, Keynes made the mistake of arguing that the unemployment which he saw in the 1930s was caused by a simple insufficiency of overall demand when it was in fact caused by the failure of relative prices to adjust to changed circumstances. Attempts to use monetary means to deal with this problem by increasing overall demand would have the exactly opposite effect to that intended. For by increasing in the short run the demand in the economy, the government would be luring more and more workers into a pattern of employment and wages which is not sustainable in the long run.

Attempts to use expansionary means to cure unemployment can only, in Professor Hayek's view, work if the real wages of those in a sector for which demand has fallen are adjusted downwards through a failure of money wages in that sector to keep pace with rising prices. But in a world where society does not have the will or the means to force unions to accept money wage cuts it is most unlikely that it will be possible to bring about real wage cuts through inflation. As unions push for higher wage settlements to recoup the ground governments are forced to induce ever faster inflation. This is the "tiger by the tail" which is pulling the western economies and the societies which rest on them to destruction.

The picture is in many ways a much gloomier one than that which emerges from the writings of some monetarist economists. At the heart of the more optimistic versions of monetarism is the belief that if money supply is kept in check then the market mechanism will work and work relatively painlessly. In order to ease the transition from a period of high inflation to a stable form of economic activity, a gradual rather than a sudden deceleration of the growth of money supply is urged. Professor Hayek is sceptical of this part of the prescription and of the optimism which underlies it. A concern not to inflate the money supply is crucial; but he at least has no doubts that this alone will not solve the problem.

The role of the trade unions as monoplies interfering with the working of the labour market lies at the very heart of his thinking. Correct monetary policy can ensure that inflation is not made worse from monetary causes, but it cannot ensure that inflation is abolished. Only a willingness to force unions to become governed by the rule of law can do that, he argues.

The question of whether, in a free society, monetary policy can by itself be sufficient to prevent inflation and produce stability must to some extent still be an open one. What is clear is that the warnings of the dangers to which an uncritical acceptance of Keynes's teaching were exposing us voiced by Professor Hayek some forty years ago have been proved right. They were forgotten for too long.

In May 1978, Hayek was the subject of this leader in *The Times* of London, heralding 'the age of Hayek' after 'the age of Keynes'.

The sea-change in politics in the UK

Margaret Thatcher, the leader of the Conservative Party 1975–1990 and Prime Minister of the United Kingdom 1979–1990, was unstinting in her praise of Hayek. On his 90th birthday in 1989, she wrote to him, saying: 'The leadership and inspiration that your work and thinking gave us were absolutely crucial and we owe you a great debt'.

PORTRAIT

HOW THE NEW RIGHT GAINED ITS POWER

US CONNECTION

Friedman

The man who is Mrs. Thatcher's bedside reading

Friedrich Von Hayek

HUGH HEBERT meets the veteran economist who counts Mrs Thatcher as a convert

The guru who still aims to get it right

Guardian 28.4.79

"YOU MUST sit on my right — I am deaf on the left. It is a source of many political jokes." That's not surprising. Professor Friedrich von Hayek

TIME

MAY 14, 1979

$1.25

Teddy Kennedy Carter Says "Baloney"

Britain's Fighting Lady

Prime Minister Margaret Thatcher

10 DOWNING STREET

LONDON SW1A 2AA

THE PRIME MINISTER

5 May 1989

Dear Professor Hayek,

It is with the greatest pleasure that I send you my warm congratulations on the occasion of your 90th Birthday. It is a tremendous milestone, especially for someone who has achieved so much.

It is ten years this week since I was privileged to become Prime Minister. Many people have been most kind in their comments on what the Government has been able to achieve in that time. There is of course still so much to do. But none of it would have been possible without the values and beliefs to set us on the right road and provide the right sense of direction. The leadership and inspiration that your work and thinking gave us were absolutely crucial; and we owe you a great debt.

With every good wish

Yours sincerely

Margaret Thatcher

Professor Friedrich August Hayek, C.H.

The sea-change in politics in the USA

Meanwhile, in the United States of America, the political change saw Ronald Reagan elected President in 1980 and again in 1984.

In 1991, in recognition of his steadfast intellectual contribution to the defence of freedom, the United States presented Hayek with the Presidential Medal of Freedom, the country's highest civilian award.

The citation read: 'Friedrich August von Hayek has done more than any thinker of our age to explore the promise and contours of liberty. He grew up in the shadow of Hitler's tyranny and devoted himself at an early age to the nurture of institutions that preserve and expand freedom.... *The Road to Serfdom* still thrills readers everywhere, and his subsequent works inspire people throughout the world.... Professor von Hayek has revolutionized the world's intellectual and political life. Future generations will read his works with the same sense of discovery and awe that inspire us today.'

Hayek being received by President Reagan in the White House 1983.

Dr Laurence Hayek receives the award on behalf of his father from President George Bush and Mrs Bush, 1991.

In the headlines

OCTOBER 1, 1979

ONE DOLLAR FIFTY CENTS

Forbes

FRIEDRICH A. HAYEK

THE REVOLT AGAINST KEY

Hayek features as a cover story in the prestigious *Forbes* magazine, October, 1979.

The New York Times 7-5-79

New Vogue for Critic of Keynes

Von Hayek Still Abhors Big Government

By JOHN M. GEDDES

Special to The New York Times

FREIBURG, West Germany, May 6 — Friedrich A. von Hayek, one of the strongest, and at times one of the few, economic voices advocating the reduction of government's role in the economy, is finally finding himself in the position of a Cassandra who suddenly that discovers people are listening.

On the eve of his 80th birthday, the Nobel laureate is back in fashion.

Margaret Thatcher, Britain's newly elected Prime Minister, is said to be a strong supporter of his views. And although Professor von Hayek labels the Tory win in Britain a "passing phenomenon," he obviously relishes the ... his economic philosophy that the ...

Von Hayek at a Glance

A leading conservative economist of this century, long an enemy of the welfare state. Ironically, in 1974, he shared the Nobel Memorial Prize in Economics with Gunnar Myrdal, whose views in ... cases are in direct contradiction to ... von Hayek ... Became

With the Thatcher Victory

Von Hayek: Cassandra Finds a New Audience

By John M. Geddes

FREIBURG, West Germany (NYT) — Friedrich von Hayek, one of the strongest, and at times one of the few, economic voices advocating the reduction of the government's role in the economy, is finally finding himself in the position of a Cassandra who suddenly discovers that people are listening.

The Nobel laureate — who just turned 80 — is back in fashion.

Margaret Thatcher, Britain's newly elected prime minister, is said to be a strong supporter of his views. And although Prof. von Hayek labels the Tory victory in Britain as a "passing phenomenon," he obviously relishes the lift for his economic philosophy that the victory signifies.

"It gives me great satisf... ings are at last ... "Whe...

He did pioneering work on business cycles, price theory and production. One of his prime conclusions was that excess expansion in the public sector, government deficit spending and expansion in the money supply of central banks were main causes of economic problems.

In 1944, by then a British subject, Prof. von Hayek wrote the impassioned "Road to Serfdom." In it he argued that movements toward socialism, toward centralized planning, involved the loss of personal freedom and eventual totalitarianism. The arguments, primarily directed at the socialist roots of Nazism in Germany, were, in effect, a broadside at Keynesian economic tho...

...nt has to decide how ...ed, or how many buses ...mines to operate or at ...e sold, these decisions ...inciples or settled for ...e wrote. "In the end, ... to decide whose in... ...nd these views must

... British economist ... theory advocating ...ment during a re... ...loyment, sympa... ...k's book, but he ...le wrote to Prof. ...y that what we ... less planning.

Friedrich von Hayek

services outside the economy, which the market cannot supply.

The current drive toward a welfare state to create a just economy, he feels "is nonsense. Only individual action can be just." As the distribution of health or strength or beauty cannot be made a matter of justice, so the distribution of things cannot be made just, he argues.

"Present policies are making things worse and ...

INTERNATIONAL HERALD TRIBUNE, SATURDAY-SUNDAY, DECEMBER 18-19, 1982

Friedrich Hayek

Grand Guru of Conservative Economics, Now 83, Still Bites Off All That He Can Chew

By Henry Allen

Washington Post Service

WASHINGTON — "Snuff," says Friedrich Hayek. He is a 1974 Nobel laureate in economics. He is 83, a prize old bull of conservative intellectuals and the unwavering foe of the rationalist scientific mindset of most of the famous minds of our century. He is sitting in an office at the Heritage Foundation, and he is opening a leather snuff box.

"I buy it in London at a shop that has 36 varieties of it. This is called Dr. James Robinson Judge's Mixture."

He deposits a pinch of brown powder in the hollow on the back ... behind his thumb. He ...

"Keynes never convinced me," says Hayek with the ease of a man who has weathered a a lifetime of astonished protests. "He was a relatively sensible man. Unfortunately, he died in 1946, and so it was left to his very orthodox pupils, who out-Keynesed Keynes, which dominated the next 30 years, and so we have been living through a period in which Keynes became a sort of saint and any critic of Keynes became a third-class citizen.

"Now, they've discovered that Keynes was wrong," says this former third-class citizen with a dry little laugh. "And my reputation is reviving."

He is much in demand. He flew to Washington after a conference ...

James M. Thresher, The Washington Post

Hayek, the answer man: "Keynes never convinced me."

particular metal, I think, we've yet to be in danger of running out of anything. The constant trend is that the prices of raw materials fall and the price of labor rises."

He fixes his listener with a stare that implies that the point is proved.

... in Hayekian terms, it is ...

night, the man returned and called me at the YMCA, where I was staying. I've regretted ever since not being able to say that I'd worked as a dishwasher."

The next year, he returned to Austria, where he already had become a disciple of Viennese economist Ludwig von Mises, who said socialism wasn't necessarily a ... intel...

Road to Serfdom," that he began his big slide into obscurity.

The book introduced him to a popular audience as a political theorist. It was an attack on the notion of central planning, saying that it inevitably leads to a loss of freedom. It was published by the University of Chicago Press, certainly respectable enough in academic circles. But the publishers sold it to Reader's Digest. Appearing there certified him as hopeless to the American intellectual establishment.

"It was received better in England, where they'd had a closer look at socialism, and disillusionment was already beginning to set in. George Orwell read it and reviewed it before he wrote 'Animal Farm' and '1984.'"

After 40 years at sea, he has his Nobel, he has foundations sending him around the world, he sees the great Keynesian machine sputtering and shorting out. He has a wife (his second) and two children, both scientists in England. His teaching career has included the University of Chicago from 1950 to 1962, and the University of Freiburg, where he is emeritus, but still lectures.

"I gave a series of lectures recently and not one student protested. Five years ago they would all have been. Now they agree ... however, that ...

Honours bestowed on Hayek in the 1980s

1980: Meeting the Pope with other Laureates, to discuss how to bridge the gap between religion and science.

1982: Member of the Order 'Pour le Mérite für Wissenschaften und Künste', Federal Republic of Germany.

THE LONDON SCHOOL OF ECONOMICS AND POLITICAL SCIENCE
Houghton Street, London WC2A 2AE

The Director
requests the pleasure of your company
at a lecture to be given by

PROFESSOR F. A. HAYEK
on
THE FLOW OF GOODS AND SERVICES

on the occasion of the golden jubilee
of his first lecture at L.S.E.

at 5 p.m. on Tuesday 27 January 1981
and thereafter at an informal reception
in the Vera Anstey Room

R.S.V.P. on enclosed card

1981: A memorial plaque is presented by Professor Ralf Dahrendorf, Director of the LSE, on the 50th anniversary of Hayek's 1931 lecture on 'Prices and Production'.

1984: Companion of Honour, Great Britain, a personal award of Queen Elizabeth II.

Happy times at Vent in the Austrian Tyrol in 1984 preceded the publication of Hayek's last book.

In 1988, now approaching 90 and drawing on a lifetime's experience of research, writing and lecturing, Hayek's last book, *The Fatal Conceit: The Errors of Socialism*, was published jointly by the University of Chicago Press and Routledge in the United Kingdom.

In his introduction to the book, Hayek is forthright:

'This book argues that our civilisation depends, not only for its origin but also for its preservation, on what can be precisely described only as the extended order of human cooperation, an order more commonly, if somewhat misleadingly, known as capitalism. To understand our civilisation, one must appreciate that the extended order resulted not from human design and intention but spontaneously.... This process is perhaps the least appreciated facet of human evolution.

'Socialists take a different view of these matters. They not only differ in their conclusions, they see the facts differently. That socialists are wrong about the facts *is crucial to my argument.... The dispute between the market order and socialism is no less than a matter of survival.'*

'Remarkably, his new book is as passionate and disputatious as anything he has written', commented *The Economist*. 'As well as adding up to a powerful manifesto against socialism, it is a fully accessible account of many of the main strands of Mr Hayek's thinking.... One of the outstanding political philosophers of this century has written a concise summation of his work: Hayek for everyman. It deserves to be read'.[11]

The political revolution in Eastern Europe

Hayek's ideas, developed over more than half a century as the 20th century's greatest champion of economic liberalism, were to have far-reaching effects not just in the UK and in the United States, but in many countries behind the Iron Curtain.

Reflecting some 30 years after *The Road to Serfdom* was published, he wrote in the Preface to the 1976 edition:

'Where I now feel I was wrong in this book is chiefly in that I rather understressed the significance of the experience of communism in Russia – a fault which is perhaps pardonable when it is remembered that when I wrote, Russia was our war-time ally.... Quite generally, further study of the contemporary trends of thought and institutions has, if anything, increased my alarm and concern. And both the influence of socialist ideas and the naive trust in the good intentions of the holders of totalitarian power have markedly increased since I wrote this book.'

But the holders of totalitarian power could not hold down the pressure for ever. In Eastern Europe and even in the Soviet Union itself, a new era was beginning. Seventy years on from the Russian Revolution, the Soviet Union was facing bankruptcy. It was getting harder and harder to keep control of the neighbouring countries as a new self-assertiveness grew. Mikhail Gorbachev became General Secretary of the Communist Party of the Soviet Union in 1985 promising openness and reform.

People took him at his word. In far corners of the communist world, Hayek's works were being translated and read. The new openness showed only too clearly the organised deficiencies of central planning, as Hayek had warned. Meanwhile the reformist policy of non-intervention simply accelerated the independence movements in Eastern Europe and the Soviet Republics as the Soviet Union under Gorbachev 'lost the will and means to coerce its subjects.'[12].

Hayek's challenge to the *'fatal conceit'* of collectivism was to be made real at the very end of the 1980s when the institutions of totalitarianism and central planning were torn down. This extraordinary cascade of events was seen by millions around the globe on television. Collectivism had failed. Communism had collapsed from within. The Cold War was over. A new age of reconstruction was beginning.

Mikhail Gorbachev, General Secretary of the Communist Party from 1985 and President of the Russian Federation, 1989–1991.

Below: Some of Hayek's work, translated into Russian.

The rediscovery of freedom

Forbes

Nearly half a century ago, most of the smart people sneered when Friedrich Hayek published "The Road to Serfdom." The world was wrong and Hayek right. Now almost 90, he has written a new book, but this time no one is sneering.

The world turns

The Economist

From Marx to the market

Evening Standard

Czechoslovakia halted as the nation goes on strike

THE BELLS RING FOR FREEDOM

The Daily Telegraph

300,000 march on the Kremlin to demand democracy

Gorbachev ready for showdown

Soviet communist monopoly may go

BERLIN

nach dem 9. November 1989

The map of Europe had to be redrawn as communism was rejected in East Germany, Czechoslovakia, Poland, Romania, Hungary and the Baltic States and finally in the Soviet Union itself.

People in the West may have understandably felt a sense of triumph as the governments of the command economies behind the Iron Curtain suddenly collapsed. But Hayek reminds us that it takes more than political change to secure the long term constitution of a liberal society. The price of liberty everywhere is not only eternal vigilance but constant, persuasive debate.

In the headlines worldwide

Seite 18 — HANDELSBLATT

Zeitgeschichte

Nr. 87 / Montag, 8. 5.

Der Nobelpreisträger für Wirtschaftswissenschaften und hochgeachtete Sozialphilosoph Friedrich August von Hayek feiert heute seinen 90. Geburtstag

Die Prinzipien einer freiheitlichen Gesellschaft verwirklichen

Von CHRISTIAN WATRIN

Sein erster Aufsatz erschien vor 65, sein erstes Buch vor 60 Jahren. Seitdem ist der Strom seiner Publikationen auf den Gebieten der theoretischen Nationalökonomie, der Rechtstheorie, der Sozialphilosophie, der Psychologie und der Geistesgeschichte nicht versiegt. Sein jüngstes Buch — zugleich der erste Band einer auf 24 Bände angelegten Ausgabe seines Gesamtwerkes - erschien vor wenigen Monaten. Es trägt den Titel „The Fatal Conceit. The Errors of Socialism" („Die fatale Einbildung"). Es handelt sich um die gründliche Auseinandersetzung mit den intellektuellen Irrtümern des Sozialismus, dem Irrglauben, daß der Mensch seine Welt nach

zur Geld- und Konjunktur... aber auch ...

Geldes und sein keynesianische aber auch geg... schen Monetar... Dogmengeschic... Darüber hina... eks Bestrebun... Ideen zum Du... fen. Beachtung ... te Londoner In... Affairs mitbegr... dendes Verdie... öffentlichen M... tannien zugun... schaft zukomm... dungs- und ... Mont Pelerin S... tionalen Verei... lich orientierte ... und Sozialwiss...

Importante lançamento de Hayek

„Hinir ríku hafa gefið hinum fátæku..."

— segði hinn þekkti nóbelsverðlaunahafi Friedrich A. Hayek

1981年(昭和56年) 10月23日 (金曜日)

De Telegraaf

Prof. Hayek en de weg naar moderne slavernij

J. G. HEITINK

AMSTERDAM, zaterdag

In deze tijd van economische en politieke verwarring is een boekje verschenen dat geschreven is door prof. dr. F. A. Hayek, „De weg naar Moderne Slavernij", „Waarheen leidt de huidige economische ..."

Een voorbeeld van Hayeks betoog... zijn de beslissingen over investeringen. „Onze generatie heeft vergeten dat eigen domsrechten de belangrijkste garantie voor ieders vrijheid vormen. Alleen wanneer de zeggenschap over de bedrijfsinvesteringen verspreid blijft over vele onafhankelijke personen, krijgen we als mens de kans over ons eigen doen en laten te beslissen."

Friedrich August v. Hayek (geb. 1899)

Der Erneuerer der Sozialphilosophie

Friedrich August von Hayek gehört zu den wenigen Denkern der Gegenwart, deren Wirken weltweite Anerkennung gefunden hat. Als einer der ersten erhielt er 1974 den Nobelpreis für Wirtschaftswissenschaft für seine liberale Wirtschaftstheorie ...

F. A. Hayek: Eine seiner Festlegungen ...

PORTRAIT

■ Das Gespräch ■

F. A. VON HAYEK

„Ungleichheit ist nötig"

Friedrich August von Hayek, 81, gilt als der bedeutendste Theoretiker der liberalen Gesellschaft dieses Jahrhunderts. *Wirtschaftswoche*-Redakteur Stefan Baron sprach mit dem in Österreich geborenen Nobelpreisträger für Ökonomie, der heute in Freiburg im Breisgau lebt, aber einen britischen Paß besitzt, über soziale Gerechtigkeit und Probleme des modernen Wohlfahrtsstaa...

■ Professor Hayek, immer mehr Bürger glauben, daß der Sozialstaat seine Grenzen erreicht hat. Sie haben schon 1944 in Ihrem Buch „Der Weg zur Knechtschaft" ... Stimme erhoben. ...

Hayek: Ja, zweifellos. Die Menschen beginnen zu erkennen, daß nur etwas verteilt werden kann, was auch vorhanden ist und daß, wenn man es so wenig verlockend macht, etwas zu produzieren wie heute, das Vorhandene immer langsamer wächst und schließlich sogar abnimmt.

■ Wie konnte eine derartige ökonomische Binsenweisheit so lange übersehen werden?

Hayek: Was wir heute ...

THE GURU WHO CAME IN FROM THE COLD

Friedrich Hayek, the high priest of the free market, has had a major influence on Mrs Thatcher's economic strategy but he still believes that she is tackling inflation too slowly, writes Eamonn Fingleton

The liberal teacher who turned his back on old age to lead world thought

Professor F. A. Hayek yesterday spent his 78th birthday travelling from Guatemala, where he had been at the Universidad Francisco Marroquin, to New York, on his way to Cornell University. During this month he will also ... Pennsylvania State College ... a conference on Cognition ... the Symbolic Processes) ... California.

In the last few months ... 1976 he was in New Zealand, Australia and Japan. The day he was in London be... a conference of economists ... Amsterdam which discu... social justice, the rule of ... and the control of money ... the Biennial Internati...

returning to economics. He was in full vigour after an interregnum of indifferent health and spiritual doldrums in the early 70s. Even his language was light-hearted: "I saw old age and did not like it; so I have returned ..."

remained a British citizen ever since. In Austria, where he was born, and in Germany, where he now teaches, he is legally a foreigner. He visits Britain regularly to see his ...

Arnold Plant spoke of the qualities he displayed in their early days at the LSE: "I can testify from personal experience to the ..."

ments. Hayek began as an economist and turned to political science, the theory of law, philosophy, ethics and even psychology. He has lately returned to economics with the radical proposition that economists, and he among them, have been wrong to regard the supply of money as a necessary monopoly of government. He has come full circle after a range of contemplation not ... ignored because it is before ...

that Keynes would himself have revised his judgments.) But in the circumstances Hayek may have been right to turn his attention to more enduring themes.

Hayek's relatively neglected early and middle writings on the theory of capital, the critique of "scientism", the error of applying to the social sciences the supposed methodology of the natural ... and inflation. Whatever differences others ...

Nostra stimolante intervista con un "mostro sacro" dell'economia: il premio Nobel von Hayek

IL GUAIO DEL TERZO MONDO? IL SOCIALISMO

di ALDO GIOBBIO

L'Occidente è un razionalismo astratto e velleitario che per imporre i suoi schemi alla ... le scuole nelle quali hanno studiato, questa mentalità ha ... Un ritorno alle idee liber...

The challenges remain, although Friedrich Hayek is no longer here to guide us. He died in Freiburg im Breisgau, in the newly unified Germany, on 23 March 1992 in his 93rd year.

Writing on Hayek's death in *The Guardian* Meghnad (Lord) Desai characterised him as: 'the only serious rival to Keynes as the most influential economist of the Twentieth Century. But his reputation is based more on his writings on political philosophy… in which he laid the foundations of the theory of a liberal order, than to his earlier complex and technical work on money, business cycle and capital theory for which he was awarded the Nobel Prize….'[13]

On a more personal level, Desai noted that Hayek 'remained to the end of his days aloof from the merely popular, a man of grace and infinite politeness'. And indeed those who met Hayek invariably point out his intellectual rigour even in informal debate. As Eamonn Butler wrote of him: 'Hayek's manners both in print and in person were impeccable… he was never one to build his own empire, (rather) diverting resources and encouragement to his research students, and treating the results of his research as if they were common property… those who knew him agreed that in his writings and in person he approached as near to the idea of the liberal scholar as perhaps human frailty will admit.'[14]

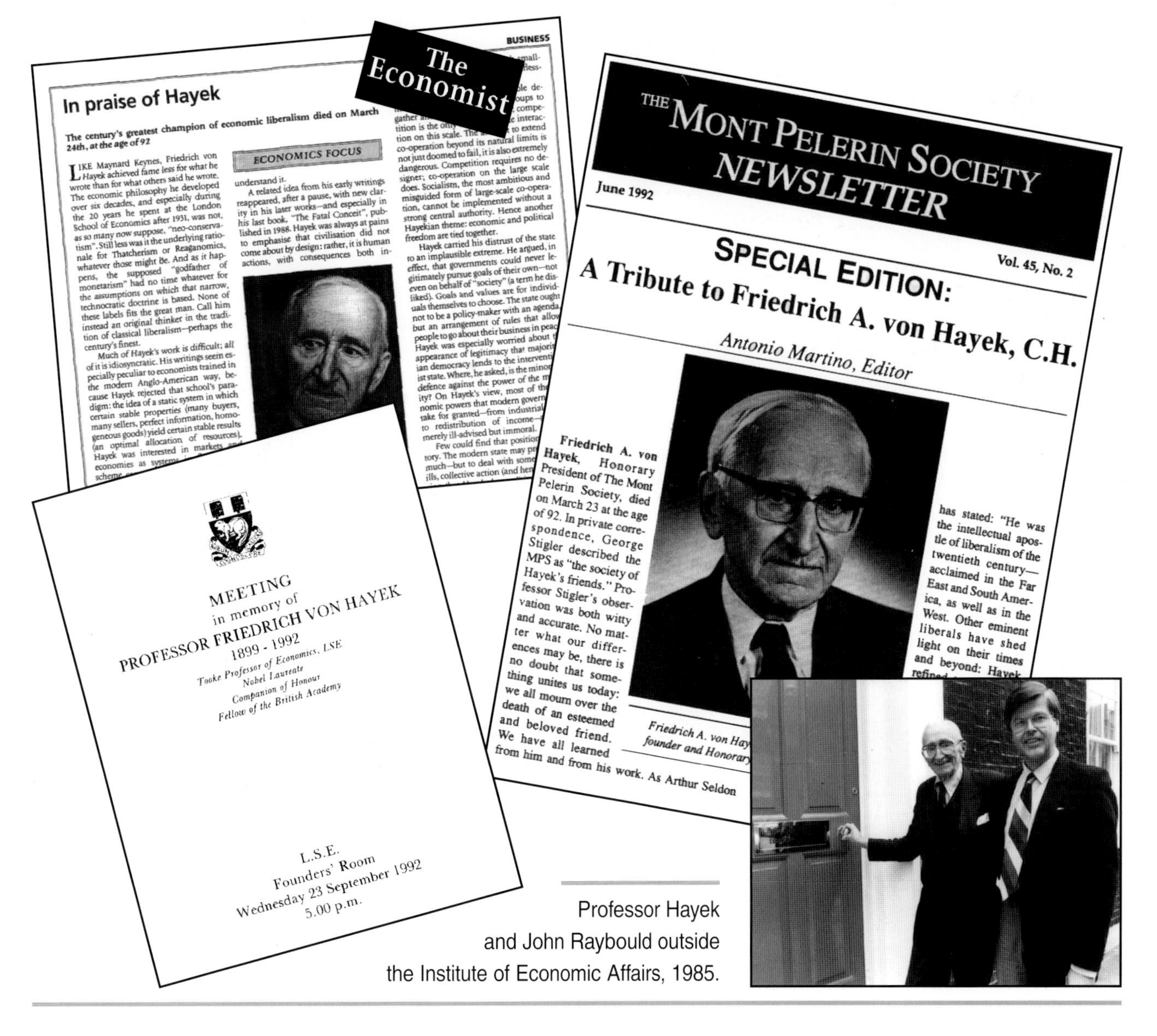

The Economist

BUSINESS

In praise of Hayek

The century's greatest champion of economic liberalism died on March 24th, at the age of 92

ECONOMICS FOCUS

LIKE Maynard Keynes, Friedrich von Hayek achieved fame less for what he wrote than for what others said he wrote. The economic philosophy he developed over six decades, and especially during the 20 years he spent at the London School of Economics after 1931, was not, as so many now suppose, "neo-conservatism". Still less was it the underlying rationale for Thatcherism or Reaganomics, whatever those might be. And as it happens, the supposed "godfather of monetarism" had no time whatever for the assumptions on which that narrow, technocratic doctrine is based. None of these labels fits the great man. Call him instead an original thinker in the tradition of classical liberalism—perhaps the century's finest.

Much of Hayek's work is difficult; all of it is idiosyncratic. His writings seem especially peculiar to economists trained in the modern Anglo-American way, because Hayek rejected that school's paradigm: the idea of a static system in which certain stable properties (many buyers, many sellers, perfect information, homogeneous goods) yield certain stable results (an optimal allocation of resources). Hayek was interested in markets and economies as systems …

understand it.

A related idea from his early writings reappeared, after a pause, with new clarity in his later works—and especially in his last book, "The Fatal Conceit", published in 1988. Hayek was always at pains to emphasise that civilisation did not come about by design: rather, it is human actions, with consequences both in-

… tition is the only … to extend co-operation on this scale. The … co-operation beyond its natural limits is not just doomed to fail, it is also extremely dangerous. Competition requires no designer; co-operation on the large scale does. Socialism, the most ambitious and misguided form of large-scale co-operation, cannot be implemented without a strong central authority. Hence another Hayekian theme: economic and political freedom are tied together.

Hayek carried his distrust of the state to an implausible extreme. He argued, in effect, that governments could never legitimately pursue goals of their own—not even on behalf of "society" (a term he disliked). Goals and values are for individuals themselves to choose. The state ought not to be a policy-maker with an agenda, but an arrangement of rules that allow people to go about their business in peace. Hayek was especially worried about the appearance of legitimacy that majoritarian democracy lends to the interventionist state. Where, he asked, is the minority's defence against the power of the majority? On Hayek's view, most of the economic powers that modern governments take for granted—from industrial policy to redistribution of income—are not merely ill-advised but immoral.

Few could find that position … tory. The modern state may … much—but to deal with some … ills, collective action (and hence …

THE MONT PELERIN SOCIETY NEWSLETTER

June 1992

Vol. 45, No. 2

SPECIAL EDITION:

A Tribute to Friedrich A. von Hayek, C.H.

Antonio Martino, Editor

Friedrich A. von Hayek, Honorary President of The Mont Pelerin Society, died on March 23 at the age of 92. In private correspondence, George Stigler described the MPS as "the society of Hayek's friends." Professor Stigler's observation was both witty and accurate. No matter what our differences may be, there is no doubt that something unites us today: we all mourn over the death of an esteemed and beloved friend. We have all learned from him and from his work. As Arthur Seldon has stated: "He was the intellectual apostle of liberalism of the twentieth century—acclaimed in the Far East and South America, as well as in the West. Other eminent liberals have shed light on their times and beyond: Hayek refined …

Friedrich A. von Hay… founder and Honorary …

MEETING
in memory of
PROFESSOR FRIEDRICH VON HAYEK
1899 - 1992
Tooke Professor of Economics, LSE
Nobel Laureate
Companion of Honour
Fellow of the British Academy

L.S.E.
Founders' Room
Wednesday 23 September 1992
5.00 p.m.

Professor Hayek and John Raybould outside the Institute of Economic Affairs, 1985.

Professor F. A. Hayek (1899–1992), photographed by Marie Gray.

Notes

Section 1: Early years in Imperial Vienna, 1899–1914

1. Hayek's *Tribute to Mises*, New York, 7 March 1956
2. Sereny (1985), interview in *The Times*
3. Kresge and Wenar (eds) (1994), p. 39
4. *Ibid.*, pp. 39, 40
5. *Ibid.*, p. 40
6. Sereny (1985), interview in *The Times*
7. Kresge and Wenar (eds) (1994), pp. 42, 43
8. *Ibid.*, p. 42
9. 'Remembering my Cousin, Ludwig Wittgenstein', *Encounter* August (1977), p. 21
10. Kresge and Wenar (eds) (1994), p. 47

Section 2: World war and revolution, 1914–1918

1. Kresge and Wenar (eds) (1994), p. 45
2. *Ibid.*, pp. 47, 48
3. Hayek (1944), pp. 80, 81
4. O'Sullivan and Raybould (1985), p. 12

Section 3: At the University of Vienna, 1918–1926

1. O'Sullivan and Raybould (1985), p. 12
2. 'The Economics of the 1920s as seen from Vienna', University of Chicago lecture, October 1963, first published in *Collected Works* vol. 4, p. 20
3. *Ibid.*, p. 22
4. *Ibid.*, pp. 22, 23
5. *Ibid.*, p. 23
6. Hayek, in a letter published 20 January 1945 in the English magazine *Time and Tide* wrote: *'It is a little hard on an ex-Austrian to be told that he does not know what misery is. I wonder how many members of the Left Intelligentsia who are so ready with that taunt have ever experienced the degree of poverty and hunger which was our common lot when I was a student in Vienna from 1918 to 1921?'*
7. 'The Economics of the 1920s as seen from Vienna' *Collected Works* vol. 4, pp. 25, 26 (Gottfried Haberler later became Professor of Economics at Harvard University)
8. O'Sullivan and Raybould (1985), p. 14
9. Hayek's Foreword to the 1981 reprint of von Mises's *Socialism: An Economic and Sociological Analysis*, Indianapolis: Liberty Classics
10. Hayek's *Tribute to Mises,* New York, 7 March 1956
11. *Collected Works* vol. 4, p. 35. Joseph Schumpeter (1883–1950) was the first Austrian exchange Professor of Economics at Columbia University 1913–1914; from 1932 to 1950 he taught at Harvard. He had studied in Vienna under Böhm-Bawerk but later moved away professionally from the Austrian School. Hayek valued the introduc-

tions Schumpeter gave him to the great American economists in the early 1920s, commenting that they were *'veritable ambassadorial letters, so large in size that I had to have a special folder made to get them uncrumpled to their destinations'*. At the time of writing the letters, Schumpeter was president of a bank in Vienna (which collapsed in 1924).

12. Kresge and Wenar (eds) (1994), pp. 88, 89
13. Butler (1983), p. 3

Section 4: The rising Austrian economist, 1927–1931

1. 'The Economics of the 1920s as seen from Vienna', *Collected Works* vol. 4, p. 36
2. Margit von Mises (1976), p. 135
3. Kresge and Wenar (eds) (1994), p. 89

Section 5: At the London School of Economics, 1931-1950

1. Skidelsky (1992), p. 456
2. Robbins (1971), p. 127
3. 'The Economics of the 1930s as seen from London', University of Chicago lecture, October 1963, first published in *Collected Works* vol. 9, pp. 52, 55
4. Robbins (1971), pp. 127, 129
5. Coase (1994), p. 209
6. Kresge and Wenar (eds) (1994), p. 81
7. Robbins (1971), p. 131
8. Galbraith (1992), Obituary of Hayek in *The Observer*, 29 March
9. Dahrendorf (1995), p. 211
10. 'The Economics of the 1930s as seen from London' *Collected Works* vol. 9, p. 49
11. *Economica*, 11(34), November 1931, pp. 387–397
12. 'Personal Recollections of Keynes and the "Keynesian Revolution"', *Oriental Economist*, 34 (663), January 1966, pp. 78–80, reprinted in *Collected Works* vol. 9, p. 241
13. O'Sullivan and Raybould (1985), p. 24
14. Cockett (1995), p. 53
15. Kresge and Wenar (eds) (1994), p. 78
16. *Ibid.*, p. 78
17. Seldon (1992), Obituary of Hayek in *The Independent*, 25 March
18. Kresge and Wenar (eds) (1994), p. 136
19. Shackle (1981), p. 242. George Shackle (1903–1992) obtained his doctorate in economics in 1937 from London University, where Hayek was his superviser. He held a chair at Liverpool University for many years.

Section 6: Hayek writes The Road to Serfdom in wartime Cambridge

1. Kresge and Wenar (eds) (1994), p. 98
2. *Ibid.*, p. 99
3. Preface to the 1956 edition of *The Road to Serfdom*
4. Harrod (1951), pp. 435, 436. Roy Harrod describes Keynes's controversy with Hayek in the 1930s as a 'sharp one', adding 'The latter, newly arrived in England, found himself terribly buffeted at this period'. Hayek's review of Harrod's book (in the *Journal of Modern History*, 24(2), June 1952) contains the following revealing passage:

   'Whatever one may think of Keynes as an economist, nobody who knew him will deny that he was one of the outstanding Englishmen of his generation. Indeed, the magnitude of his influence as an economist is probably at least as much due to the impressiveness of the man, the universality of his interests, and the power and persuasive charm of his personality as to the originality or theoretical soundness of his contribution to economics. He owed his success largely to a rare combination of brilliance and quickness of mind with a mastery of the English language ... and ... a voice of bewitching persuasiveness. As a scholar he was incisive rather than profound and thorough, guided by strong intuition which would make him try to prove the same point again and again by different routes.'

   In contrast to his own years of study of economics, Hayek continued, *'Keynes was able to master the essential outlines of a new subject in a remarkably short time; indeed, he seems to have turned himself into an economist, after a university course in mathematics, in the course of little more than two years filled with many other activities. The result of this, however, was that the scope of his knowledge remained always not only somewhat insular but distinctly "Cambridge"'.*

5. Milton Friedman's Introduction to the 50th anniversary edition of *The Road to Serfdom* (1994) Chicago: University of Chicago Press, p. xviii
6. Kresge and Wenar (eds) (1994), p. 104
7. Margit von Mises (1976), p. 107
8. Kresge and Wenar (eds) (1994), p. 103
9. Quoted in Hartwell (1995), pp. 23, 24
10. Sereny (1985), interview in *The Times*
11. Kresge and Wenar (eds) (1994), p. 106
12. Cockett (1995), p. 25
13. Robbins (1971), pp. 127, 128
14. Gray (1984), pp. 10, 11
15. Dahrendorf (1995), pp. 422, 423

Section 7: The rebirth of a liberal movement in Europe

1. Hartwell (1995), p. 35. Max Hartwell notes that an international meeting to discuss the revival of liberalism was held in Paris in 1938, inspired by the publication of *An Inquiry into the Principles of the Good Society* by the American columnist Walter Lippmann in 1937. Hayek, Mises and Röpke were among those who attended. A second meeting was held in 1939 but the Second World War prevented any more being held. Lippmann was later invited to the first meeting of the Mont Pèlerin Society in 1947 and became a founding member.
2. Hayek's paper 'Historians and the Future of Europe', which he read to the Political Society at King's College, Cambridge, 28 February 1944, was first published in *Studies in Philosophy, Politics and Economics*, 1967, pp. 135–147.
3. 'The Rediscovery of Freedom: Personal Recollections', Hayek's lecture at Bonn–Bad Godesberg, Germany, February 1983, was reprinted in *Collected Works* vol. 4, pp. 185–197.
4. Hayek's 'Opening Address to a Conference at Mont Pèlerin, April 1, 1947' was first published in *Studies in Philosophy, Politics and Economics*, 1967, pp. 148–159.
5. 'The Rediscovery of Freedom' *Collected Works* vol. 4, p. 192
6. Cockett (1995), p. 54
7. Hayek's paper '"Free" Enterprise and Competitive Order', which he read at the first meeting of the Mont Pèlerin Society in 1947, was first published in *Individualism and Economic Order*, 1948, pp.107–118.
8. *Economica*, August (1948), pp. 227–230
9. 'The Intellectuals and Socialism' was first published in the *University of Chicago Law Review*, 16(3), Spring 1949, pp. 417–433.
10. Hayek uses the term 'liberal' in its original 19th century British sense of limited government, private property rights and free markets, whereas in the United States it has acquired almost the opposite meaning, especially since the 1930s.

Section 8: At the University of Chicago, 1950–1962

1. Kresge and Wenar (eds) (1994), p. 126
2. *Ibid.*, p. 128
3. Sereny (1985), interview in *The Times*
4. *Criticon*, November 1974, reprinted in *Knowledge, Evolution and Society*, London: Adam Smith Institute, 1983, pp. 9–15
5. For an insider's view of this remarkable committee, see Nef (1973), pp. 185, 237, 238. Founded at Dr Nef's instigation in the early 1940s, the Committee on Social Thought at the University of Chicago was an interdisciplinary graduate department, which attracted to its faculty leading international scholars in the humanities and the social sciences as well as creative artists.

   Dr Nef recounts: 'During the 1949 Easter holidays I lectured at the extension school of the University of Birmingham. My visit to England, where I met T. S. Eliot and Friedrich Hayek in London, enabled me to make those two important appointments to the Committee on Social Thought. Hayek accepted a permanent chair he was destined to hold for almost fifteen years. The Economics Department welcomed his connection with Social Thought, although the economists had opposed his appointment in Economics four years before largely because they regarded his *Road to Serfdom* as too popular a work for a respectable scholar to perpetrate. It was all right to have him at the University of Chicago so long as he wasn't identified with the economists.' (pp. 237–238)

An indication of the milieu in which Hayek taught and 'an impression of the range and caliber of … the distinguished persons who participated either as visitors or as regular members' is given by Nef's list, which, in addition to Hayek and Eliot, included Hannah Arendt, Saul Bellow, D. J. Boorstin, Alfred Cobban, Marc Chagall, Colin Clark, Harold Innis, Frank Knight, Michael Polanyi, Maurice Powicke, Arnold Schoenberg, Igor Stravinsky, R. H. Tawney and Arnold Toynbee.

6. Obituary of the sociologist Edward Shils, a member of the Committee on Social Thought, *Daily Telegraph*, 8 February 1995.
7. Kresge and Wenar (eds) (1994), p. 126
8. Butler (1983), p. 6
9. Letwin (1976), p. 148
10. Butler (1983), p. 9
11. Kresge and Wenar (eds) (1994), p. 153
12. While at the University of Chicago in the 1950s, Hayek's international contacts were maintained by his regular attendance at the general meetings of the Mont Pèlerin Society held at Beauvallon (1951), Seelisberg (1953), Venice (1954), Berlin (1956), St Moritz (1957), Princeton (1958) and Oxford (1959)
13. Hartwell (1995), pp. 93, 94
14. Margit von Mises (1976), pp. 187, 188
15. Kresge and Wenar (eds) (1994), p. 129
16. *Ibid.*, pp. 129, 130
17. Butler (1983), p. 10
18. Kresge and Wenar (eds) (1994), pp. 135, 136
19. Margit von Mises (1976), p. 185

Section 9: The prophet in the wilderness, 1962–1974

1. Kresge and Wenar (eds) (1994), p. 131
2. 'The Rediscovery of Freedom' *Collected Works* vol. 4, pp. 189, 190
3. *Ibid.*, p. 189
4. *Ibid.*, pp. 193, 194
5. Hobsbawm (1994), pp. 176–177
6. 'The Keynes Centenary: The Austrian Critique', *The Economist*, 11 June 1983. In this feature article written nearly 40 years after Keynes's death, Hayek said that he still regarded Keynes *'as perhaps the most impressive intellectual figure I have ever encountered…. He was certainly one of the most powerful thinkers and expositors of his generation.*

   'But, paradoxical as this may sound', Hayek continued, *'he was neither a highly trained economist nor even centrally concerned with the development of economics as a science…. His main aim was always to influence current policy, and economic theory was for him simply a tool for this purpose…. The time when he had become the idol of the leftish intellectuals was in fact when in 1933 he had shocked many of his earlier admirers by his essay on "National Self-Sufficiency" in the* New Statesman and Nation *[where] he proclaimed that "The decadent international but individualistic capitalism, in the hands of which we found ourselves after the war, is not a success. It is not intelligent, it is not beautiful, it is not just, it is not virtuous – and it does not deliver the goods. In short, we dislike it and we are beginning to despise it." Later, still in the same mood, in his preface to the German translation of* The General Theory, *he frankly recommended his policy proposals as being more easily adapted to the conditions of a totalitarian state than those in which production is guided by free competition.'*
7. Fisher (1974), p. 104
8. Kresge and Wenar (eds) (1994), p. 130
9. *Ibid.*, p. 131
10. Butler (1983), p. 7

Section 10: The power of ideas, 1974–1992

1. Obituary in *The Times*, 25 March 1992
2. Butler (1983), p. 7
3. Sereny (1985), interview in *The Times*
4. Lundberg, Erik, *Nobel Lectures in Economic Science, 1969–1980* Singapore: World Scientific Publishing, 1992, pp. 173–175

5. Hayek's Nobel Memorial Lecture of 11 December 1974 was reprinted in *New Studies in Philosophy, Politics, Economics and the History of Ideas* (1978), pp. 23–34.
6. *University of Chicago Law Review*, 19(3), Spring 1949
7. 'The Rediscovery of Freedom' *Collected Works* vol. 4, p. 193
8. Butler (1985), p. 7
9. *Policy Report* Washington, DC: Cato Institute, 1983
10. 'The Rediscovery of Freedom', *Collected Works* vol. 4, pp. 185, 194
11. *The Economist*, 28 January 1989
12. Skidelsky (1995), p. 95 *et seq.*, particularly the chapter 'Why did Soviet Communism collapse?'
13. Desai, Meghnad (Lord), Obituary of Hayek in *The Guardian*, 25 March 1992
14. Butler (1985), pp. 12, 13

Sources and references

Butler, Eamonn *Hayek: His Contribution to the Political and Economic Thought of Our Time* London: Temple Smith 1983

Chronicle of the 20th Century Farnborough, Hants: Chronicle Communications 1993

Coase, R. H. 'Economics at LSE in the 1930s: A Personal View' *Atlantic Economic Journal*, 31–34, 1982

Cockett, Richard *Thinking the Unthinkable: Think-Tanks and the Economic Counter-Revolution, 1931–1983* London: HarperCollins 1995

Dahrendorf, Ralf *A History of the London School of Economics and Political Science 1895–1995* Oxford: Oxford University Press 1995

Fisher, Antony *Must History Repeat Itself?* London: Churchill Press 1974

Gamble, Andrew *Hayek: The Iron Cage of Liberty* Cambridge: Polity Press 1996

Garrison, Roger W. and Kirzner, Israel M. 'Friedrich August von Hayek' in *The New Palgrave: A Dictionary of Economics* London: Macmillan 1987

Gray, John *Hayek on Liberty* Oxford: Basil Blackwell 1984

Harrod, R. F. *The Life of John Maynard Keynes* London: Macmillan 1951

Hartwell, R. M. *A History of the Mont Pèlerin Society* Indianapolis, IN: Liberty Fund 1995

Hayek, F.A., The Collected Works

Volume 1: *The Fatal Conceit: The Errors of Socialism*, ed. W. W. Bartley III 1988

Volume 3: *The Trend of Economic Thinking: Essays on Political Economists and Economic History*, ed. W. W. Bartley III and Stephen Kresge 1991

Volume 4: *The Fortunes of Liberalism: Essays on Austrian Economics and the Ideal of Freedom*, ed. Peter G. Klein 1992

Volume 9: *Contra Keynes and Cambridge Essays, Correspondence*, ed. Bruce Caldwell, 1995

Volume 10: *Socialism and War Essays, Documents, Reviews*, ed. Bruce Caldwell 1997

Supplement: *Hayek on Hayek: An Autobiographical Dialogue*, ed. Stephen Kresge and Leif Wenar 1994

Publication of Hayek's *Collected Works* is being undertaken between Routledge in the United Kingdom and the University of Chicago Press in the United States. Some 20 volumes are anticipated, of which the above six have been published to date.

Hobsbawm, Eric *Age of Extremes: The Short Twentieth Century 1914–1991* London: Michael Joseph 1994

Letwin, Shirley Robin 'The Achievement of Friedrich A. Hayek' in Fritz Machlup (ed.) *Essays on Hayek* Hillsdale, MI: Hillsdale College Press, 1976

Mises, Margit von *My Years with Ludwig von Mises* New Rochelle, NY: Arlington House 1976

Nef, John U. *Search for Meaning: The Autobiography of a Nonconformist* Washington, DC: Public Affairs Press 1973

O'Sullivan, John and Raybould, John *Hayek: His Life and Thought*, a 78-minute video cassette interview of Professor Hayek by John O'Sullivan with a *Viewer's Guide* with transcript prepared by John Raybould. London: Institute of Economic Affairs and Video Arts 1985. This interview was televised in full on BBC2 on 20 November 1986.

Raybould, John *Hayek: A Tribute*, a slide presentation, commissioned by the Fraser Institute and the Mont Pèlerin Society and shown at the 1992 general meeting of the Society in Vancouver; subsequently released as a 24-minute video cassette. London: Adam Smith Institute and Hartfield, Sussex: Atlas Economic Research Foundation 1993

Robbins, Lionel (Lord) *Autobiography of an Economist* London: Macmillan 1971

Sereny, Gitta 'Sage of the free-thinking world', *The Times Profile: Friedrich August von Hayek* London: *The Times*, 9 May 1985

Shackle G. L. S. 'F. A. Hayek' in O'Brien, D. P. and Presley, John R. (eds) *Pioneers of Modern Economics in Britain* London: Macmillan 1981

Skidelsky, Robert *John Maynard Keynes: The Economist as Saviour 1920–1937* London: Macmillan 1992

Skidelsky, Robert *The World After Communism: A Polemic For Our Times* London: Macmillan 1995

Select bibliography

During his long career, Hayek wrote 17 books and edited or introduced a further 16. He wrote some 25 pamphlets and upwards of 250 articles that were published in collections of essays, encyclopaedias and in scholarly journals such as *Economica* and the *Economic Journal* in which he and Keynes maintained their vigorous debates in the 1930s.

Hayek wrote many 'Letters to the Editor' as well as book reviews. In later life he enthusiastically took part in interviews on the radio and on film and video. He maintained extensive correspondence with the leading intellectuals of the day as well as with the various free-market institutes and allied organizations around the globe. For over 40 years he actively participated in the meetings of the Mont Pèlerin Society, which still continues the debate and development of the ideas of liberty and provides a meeting place for those of many different nationalities who wish to be part of that development.

An extensive bibliography of Professor Hayek's works can be found in John Gray's *Hayek on Liberty* (1984, pp. 143–209), which was compiled with the assistance of Hayek himself.

Hayek's books in English

Prices and Production London: Routledge 1931 and New York: Macmillan 1932

Monetary Theory and the Trade Cycle London: Jonathan Cape 1933 and New York: Harcourt Brace 1933 (this was the English translation of Hayek's first book in Vienna *Geldtheorie und Konjunkturtheorie*, 1929)

Monetary Nationalism and International Stability Geneva: 1937

Profits, Interest and Investment and Other Essays on the Theory of Industrial Fluctuations London: Routledge 1939

The Pure Theory of Capital London: Routledge 1941 and Chicago: Chicago University Press 1941

The Road to Serfdom London: Routledge 1944 and Chicago: University of Chicago Press 1944

Individualism and Economic Order London: Routledge 1948 and Chicago: University of Chicago Press 1948

John Stuart Mill and Harriet Taylor: Their Friendship and Subsequent Marriage London: Routledge 1951 and Chicago: University of Chicago Press 1951

The Counter-Revolution of Science: Studies on the Abuse of Reason Glencoe, IL: The Free Press 1952; 2nd edn Indianapolis, IN: Liberty Press 1979

The Sensory Order: An Inquiry into the Foundations of Theoretical Psychology London: Routledge 1952 and Chicago: University of Chicago Press 1952

The Political Ideal of the Rule of Law Cairo: National Bank of Egypt 1955

The Constitution of Liberty London: Routledge 1960 and Chicago: University of Chicago Press 1960

Studies in Philosophy, Politics and Economics London: Routledge 1967 and Chicago: University of Chicago Press 1967

Law, Legislation and Liberty: A New Statement of the Liberal Principles of Justice and Political Economy London: Routledge and Chicago: University of Chicago Press
Volume I *Rules and Order* 1973
Volume II *The Mirage of Social Justice* 1976
Volume III *The Political Order of a Free People* 1979

New Studies in Philosophy, Politics, Economics and the History of Ideas London: Routledge 1978 and Chicago: University of Chicago Press 1978

Money, Capital and Fluctuations: Early Essays, edited by Roy McLoughry with an introduction by F. A. Hayek, London: Routledge, in co-operation with the London School of Economics 1984 (English translation from the original ten German articles, 1925–1936)

The Fatal Conceit: The Errors of Socialism London: Routledge 1988 and Chicago: University of Chicago Press 1989

Books edited and introduced in English

The Collected Works of Carl Menger (4 vols) London: London School of Economics 1933–1936

Collectivist Economic Planning: Critical Studies on the Possibilities of Socialism London: Routledge 1935

Capitalism and the Historians London: Routledge 1954 and Chicago: University of Chicago Press 1954

Hayek's principal short papers in English

Freedom and the Economic System Chicago: University of Chicago Press 1939

The Confusion of Language in Political Thought, With Some Suggestions for Remedying It London: Institute of Economic Affairs 1968

A Tiger by the Tail: The Keynesian Legacy of Inflation: A 40 Years' Running Commentary on Keynesianism, compiled and edited by Sudha R. Shenoy London: Institute of Economic Affairs 1972; 2nd edn 1978

Full Employment at Any Price? London: Institute of Economic Affairs 1975

Choice in Currency: A Way to Stop Inflation London: Institute of Economic Affairs 1976

Denationalisation of Money: An Analysis of the Theory and Practice of Concurrent Currencies London: Institute of Economic Affairs 1976; 3rd edn 1990

1980s Unemployment and the Unions London: Institute of Economic Affairs 1980; 2nd edn 1984

Knowledge, Evolution and Society London: Adam Smith Institute 1983

Picture acknowledgements

The author and publisher wish to thank the following individuals and organisations for providing them with the photographs and documents that made this publication possible, and for granting permission to publish them.

Dr Laurence Hayek (items on pp. 1–7, 9, 10, 12–23, 25, 26, 28, 29, 32, 35–37, 41, 43, 46, 53, 56–63, 65, 66, 68–71, 73–75, 77–82, 84, 85, 90, 91, 93–96, 101, 102, 104, 105)

Popperfoto (pp.11, 27, 30, 51)

Fabian Society (p. 11)

Estate of Margit von Mises (pp. 15, 48)

Bruce Bernstein, courtesy of Princeton University Library (p. 18)

Miss Christine Hayek (pp. 20, 34, 42)

Macmillan Publishers (p. 22, 45)

Illustrated London New Picture Library (p. 23)

Ronald H. Coase (pp. 30, 97)

Cambridgeshire Libraries (pp. 31, 38, 39, 41, 44)

London School of Economics (p. 31, 104)

Marshall Library, Faculty of Economics and Politics, University of Cambridge (pp. 31, 40)

King's College Archives (pp. 32, 33, 40, 45, 55)

Austrian National Library (p. 34)

Hulton-Getty Picture Library (pp. 42, 64, 80)

Reader's Digest (p. 48)

BBC (p. 49)

Government of Austria (pp. 52, 84))

TASS (p. 54)

University of Chicago (p. 67)

Professor William Letwin (p. 69)

The Times (p. 76)

Mont Pèlerin Society (pp. 83, 109)

Foto Vouk, Vienna (p.86)

Fritz Kern, Vienna (p. 86)

David Farrer (pp. 87, 104)

Hannes Gissurarson (p. 87)

Marie Gray (pp. 88, 110, cover)

Nobel Foundation (pp. 89, 97)

Mick Moore (p. 95, 96)

Fife Free Press (p. 96)

Eric Brodin (p. 97)

Time (p.101)

The White House (p. 102)

Forbes magazine (p. 103)

Pontificia Fotografia (p. 104)

Russian Embassy in the UK (p. 106)

The Economist (p. 109)

All other illustrations come from the author's collection